STEWARDS OF DATA

STEWARDS OF DATA

A Practical Handbook for Undergraduate Research in Engineering and Applied Sciences

ŞENAY PURZER, WEI ZAKHAROV, AND CARLA B. ZOLTOWSKI

PURDUE UNIVERSITY PRESS
WEST LAFAYETTE, INDIANA

Cataloging-in-Publication Data on file at the Library Congress.

978-1-62671-390-1 (paperback)
978-1-62671-391-8 (epdf)

Cover image: A collection of important metadata elements used in data documentation.
Created by Sarah Michelle Sewell.

CONTENTS

1

STEWARDS OF DATA: AN INTRODUCTION

1.1 SUMMARY: WHAT IS THIS BOOK ABOUT?

This book is an interdisciplinary endeavor that integrates ethical data life cycle management education into undergraduate research programs in engineering and applied sciences. Drawing on research in engineering education, data management, and ethical reasoning, the project team introduces a data stewardship framework designed to support research mentors, graduate students, and undergraduate researchers in the responsible care and management of data throughout the research process. As artificial intelligence (AI) systems increasingly rely on large, complex, and heterogeneous datasets, effective data management and ethical decision-making have become even more critical. This book responds to this growing need by foregrounding data stewardship as a foundational research skill, supporting transparent, reproducible, and ethically grounded research practices across the data life cycle.

The data stewardship framework helps researchers identify underlying disciplinary principles, apply data management processes, consider ethical implications, and articulate trade-offs involved in decision-making across the data life cycle. To move beyond abstract guidance, the book pairs the framework with a practical toolkit that includes worksheets, reflection prompts, rubrics, and mentoring conversation guides. These tools are designed to be used in authentic research settings, enabling students and mentors to translate ethical principles into everyday research practices.

Throughout the book, real-world use cases drawn from undergraduate research contexts illustrate common ethical data management challenges. These examples highlight situations such as organizing and documenting data, sharing datasets responsibly, balancing openness with privacy or intellectual property concerns, and preparing data for reuse. Together, the framework, toolkit, and examples provide readers with concrete strategies for cultivating ethical data stewardship skills and embedding them into undergraduate research experiences.

1.2 TARGET AUDIENCE: WHO IS THIS BOOK FOR?

This book is written for individuals and programs that support new researchers and analysts who work with data in engineering and applied sciences. While readers may come from different roles and disciplines, they share a common goal: enabling responsible, ethical, and effective research data management practices across the data life cycle.

UNDERGRADUATE RESEARCHERS

This guide is designed to support undergraduate researchers who are actively engaged in data-driven projects and are navigating research expectations for the first time. It provides practical strategies for planning, organizing, documenting, sharing, and preserving data, while also helping students recognize design and ethical considerations and trade-offs embedded in these activities. Rather than assuming prior expertise, the book introduces core concepts using accessible language, real-world examples, and structured reflection and includes many examples from undergraduate researchers working on projects that use machine learning models.

RESEARCH MENTORS AND INSTRUCTORS

Faculty members, graduate student mentors, and teaching assistants who supervise undergraduate research will find this book useful as a mentoring and instructional resource. It supports conversations about data practices that often remain implicit, such as documentation, file naming convention and organization, data ownership, reuse expectations, and ethical responsibility. The framework and tools can be integrated into undergraduate research courses, capstone design experiences, lab onboarding practices, mentoring conversations, and experiential learning programs.

RESEARCH PROGRAMS AND SUPPORT OFFICES

This book also speaks to undergraduate research programs, librarians, data stewards, instructional designers, archivists, and staff in compliance, IT, or sponsored research offices. These readers may use this handbook to develop training materials, workshops, or program-level interventions that align infrastructure, policy, and instruction around responsible data stewardship.

ENGINEERS AND PROFESSIONAL RESEARCHERS

Engineers and researchers in professional or industry settings will find value in revisiting foundational data management and ethical practices, particularly as expectations for transparency, reproducibility, and responsible data use continue to evolve.

POLICYMAKERS, FUNDERS, AND INSTITUTIONAL LEADERS

Leaders and decision-makers interested in advancing findable, accessible, interoperable, and reusable (FAIR) data practices will find the data stewardship framework useful for understanding how ethical data competencies can be cultivated early in the research pipeline.

In short, if you are conducting, mentoring, supporting, or shaping undergraduate research that involves data, this book is written for you. The following use cases illustrate common data life cycle and ethical challenges faced by undergraduate researchers. Each scenario highlights one dimension of data stewardship and reflects situations that arise frequently but are rarely addressed explicitly in undergraduate research settings.

SCENARIO 1: PREPARING DATA FOR CONTINUITY AND REUSE

Ben has worked as an undergraduate researcher on a smart cities project for two semesters. As he prepares to graduate in May, new undergraduate researchers will be joining the project in the fall. Ben must decide how to organize files, document data collection methods, define variable meanings, and explain analytical decisions so others can understand and continue the work. What level of documentation is sufficient? How much contextual information is necessary for reuse? What ethical responsibilities does Ben have to future team members and to the broader research community?

SCENARIO 2: DESIGN AND ETHICAL DECISION MAKING

Stephanie is developing an AI algorithm to predict user behaviors using data from a new health-monitoring technology. She must consider issues such as data consent, bias, representativeness, and potential downstream impacts of model predictions. What assumptions are embedded in the data and data management? Who might be excluded or misrepresented? How should uncertainty and limitations be documented? Stephanie's decisions during data selection, modeling, and evaluation carry ethical implications that extend beyond technical performance.

SCENARIO 3: PUBLISHING AND SHARING RESEARCH DATA

Alex is responsible for publishing datasets alongside the research group's journal articles. Alex must determine where to deposit the data, what metadata to include, how to handle sensitive or proprietary information, and how to license the data for reuse. Which repository is appropriate? How can the data be made discoverable and interpretable by others? How does Alex balance openness with ethical and legal constraints?

Together, these scenarios demonstrate that data stewardship is not a single step or checklist item, but a series of decisions made throughout the research process. This book uses such scenarios to surface implicit assumptions, illuminate trade-offs, and provide practical tools that help undergraduate researchers and those who support them navigate these challenges with intention and care.

1.3 SIGNIFICANCE: WHY READ THIS BOOK?

Data now underpin nearly every stage of the research process, from problem formulation and design to analysis, dissemination, and long-term reuse. Recognizing this shift, the National Academies of Sciences, Engineering, and Medicine

(NASEM, 2018) identifies 10 core competencies necessary for data-driven research. Notably, seven of the areas (identified with bold italics) are directly related to ethical data life cycle management. This emphasis underscores a growing recognition that responsible data practices are not peripheral skills but foundational competencies for next-generation researchers.

- Mathematical foundations
- Computational foundations
- Statistical foundations
- ***Data management and curation***
- ***Data description and visualization***
- ***Data modeling and assessment***
- ***Workflow and reproducibility***
- ***Communication and teamwork***
- ***Domain-specific considerations***
- ***Ethical problem-solving***

Yet a persistent challenge remains: How can undergraduate researchers be equipped effectively with ethical data management skills in practice? Researchers across disciplines increasingly rely on data-intensive methods, leading to rapid growth in the volume, variety, and complexity of research data. At the same time, a substantial gap is emerging between the creation of research data and its meaningful sharing and reuse. Producing scientific knowledge alone is no longer sufficient; the value of research is increasingly determined by whether its underlying data can be found, understood, trusted, and reused by others.

Policy developments over the past decade have further reinforced these expectations. The 2013 *Public Access to Federally Funded Research* memorandum from the White House Office of Science and Technology Policy directed federal funding agencies to establish requirements for public access to research articles and data. Many agencies now mandate data management plans (DMPs) as part of the grant application process. For example, since January 18, 2011, the National Science Foundation has required all proposals to include a supplementary DMP, signaling a clear expectation that research data be responsibly managed and shared. Increasingly, private foundations have adopted similar requirements.

These policies have contributed to the growth of data citation practices, data repositories, and data journals. However, public availability alone does not guarantee usability. Undergraduate researchers, in particular, often struggle to locate relevant datasets, interpret documentation, assess data quality, and determine whether data can be ethically and appropriately reused. Although contemporary digital platforms have made data sharing easier than ever, ease of sharing does not translate into ease of reuse, nor does it ensure ethical decision-making. As a result, students are increasingly expected to both consume and produce data that are findable, accessible, interoperable, and reusable (FAIR), often without explicit instruction in how to do so responsibly.

This book responds directly to this gap. Drawing on research in engineering education, data life cycle management, and ethical reasoning, this book positions undergraduate researchers not only as data users or analysts but as data stewards—individuals responsible for the long-term care, integrity, and ethical use of research data. Data stewardship extends beyond policy compliance; it requires navigating trade-offs, recognizing risks and benefits, and making informed decisions at every stage of the data life cycle. By prioritizing data stewardship early in researchers' careers, this handbook aims to strengthen not only individual research projects, but also the broader research ecosystem.

1.4 HOW SHOULD THIS PRACTICAL HANDBOOK BE USED?

This book is designed as a practical, flexible guide. Readers are encouraged to engage with the chapters and tools in ways that best align with their roles, projects, and immediate needs.

For undergraduate researchers, the book can be used alongside an active research project. The worksheets, examples, and reflective questions are intended to support decision-making at key moments across the data life cycle—from early project planning and data collection to analysis, sharing, and publication. Students may find it most useful to revisit specific sections of this book multiple times as their projects evolve.

For graduate and faculty mentors, the book can serve as a modular resource. Individual chapters, figures, or toolkit components may be integrated into undergraduate research courses, capstone design experiences, lab onboarding practices, or mentoring conversations. The framework and use cases are designed to provoke discussion about ethical trade-offs and disciplinary norms rather than prescribe a single "correct" approach.

For research program administrators, librarians, and data professionals, this handbook can support curriculum development, training workshops, and program-level efforts to embed responsible data practices into undergraduate research experiences. The data stewardship framework offers a shared vocabulary that can help align instruction, infrastructure, and policy.

Across audiences, readers are encouraged to treat this book as a living reference that supports reflection, iteration, and adaptation rather than compliance alone.

1.5 HOW DID THIS BOOK COME ABOUT?

This book emerged from several years of interdisciplinary collaboration among faculty in academic libraries, disciplinary engineering schools, and engineering education, as well as sustained engagement with undergraduate researchers and their mentors. It builds on a National Science Foundation Improving Undergraduate STEM Education (IUSE)-funded project focused on prioritizing data life cycle ethical management in undergraduate research programs in engineering and applied sciences.

The initial development began by gathering worksheets, instructional materials, and resources that the authors had created and refined through undergraduate research courses, research programs and labs, and mentoring contexts. In parallel, we reviewed existing data management workbooks and practical guides, including Briney's (2015) *Data Management for Researchers* and related toolkits, to identify both effective practices and gaps—particularly those related to ethics, decision-making, and undergraduate research contexts.

Through this process, we refined our approach around the concept of data stewardship, emphasizing not only technical data management tasks but also design and ethical considerations, future consequences, and trade-offs that arise throughout the data life cycle. This framing draws on existing literature and frameworks while extending them to more holistically and explicitly foreground responsibility, care, and context-sensitive judgment in research practice.

As the project progressed, we further refined the toolkit with a deliberate focus on accessibility for undergraduate researchers. Many core concepts, such as metadata, data citation, intellectual property, and data reuse, can be unfamiliar or overwhelming to new researchers. To address this, we adopted familiar representations, concrete examples, and use cases that reflect real undergraduate research scenarios. Feedback from undergraduate researchers played an important role in shaping the language, structure, and emphasis of the toolkit.

The result is a resource grounded in research and practice, shaped by interdisciplinary expertise, and informed by the lived experiences of undergraduate researchers.

1.6 USE OF AI IN THE WRITING OF THIS BOOK

In keeping with the values of transparency and ethical practice that underpin this book, we are explicit about our use of AI tools during the writing process. Members of the author team used AI-assisted tools to support limited and well-defined tasks, i.e. assisting with initial surface-level editing for grammar checking and clarity.

All core ideas, frameworks, interpretations, examples, and pedagogical designs presented in this book were developed by the author team. AI tools were not used to generate original scholarly arguments, conceptual frameworks, or empirical claims. Final decisions about content, structure, and wording were made by the authors, who remain fully responsible for the accuracy and integrity of the work. We view this approach as consistent with responsible data stewardship: AI tools can be valuable when used thoughtfully and transparently.

2

DATA STEWARDSHIP FRAMEWORK

2.1 SUMMARY

This chapter provides a definition of data stewardship and a framework that serves as a foundation for the other chapters that offer practical guidance to research mentors. This chapter also defines a researcher as someone who generates knowledge and situates data as the foundation of this knowledge. The role of data in research has always been critical, and this importance has grown and renewed as artificial intelligence (AI) continues to shape the practices of researchers.

2.2 A RENEWED DEFINITION OF RESEARCHER

Traditionally, a researcher has been defined as someone who conducts systematic investigations. In addition, research is also typically characterized by a cycle of data collection and data analysis. However, these definitions are incomplete without an articulation of purpose. While each discipline has different research traditions, the goal of research across all disciplines is to create new knowledge.

The generation of new knowledge begins with data. Hence, without organized data, research is not possible. In this sense, organized data serve as the foundation for research. Yet, data by themselves do not produce knowledge; they must be interpreted and contextualized through specialized research methods and theoretical perspectives. In a metaphoric epistemic engine, data are the air and fuel processed through pistons and gears, and knowledge is the power and value generated by this engine. As illustrated in Figure 2.1, the process of creating new knowledge relies on an iterative cycle in which raw data evolves into information, information evolves into knowledge, and in turn, this knowledge guides the next round of data collection and processing. The figure depicts the data–information–knowledge continuum explained through the rhetoric tool of asking why, what, and how.

A dataset is the researcher's attempt to capture the intrinsic meaning of a real event (Liew, 2007). Data are captured and stored as historical records of past events. When an event is captured and stored in the form of data, we create a resource for future information and knowledge creation. Hence, responsible data practices are required in support of the full research life cycle.

Data analysis and visualization methods enable the creation of a reconstructed picture of data. By processing the data into useful pieces of information, future insights can be gained. Such analysis and visualizations become information that guides decision-making and problem-solving. This information becomes knowledge after human interpretation, synthesis, and judgment. When this new piece of knowledge adds to the existing body of knowledge, the process is called the *culmination of expertise*. Hence, when a researcher creates new knowledge, it comes with the responsibility of caring for the data created, information extracted, and knowledge generated.

As data-intensive research evolves, the definition of the researcher should evolve, too, with a stronger emphasis on data. One of the most important duties of a researcher is the ethical management of data (Frugoli et al., 2010). Despite

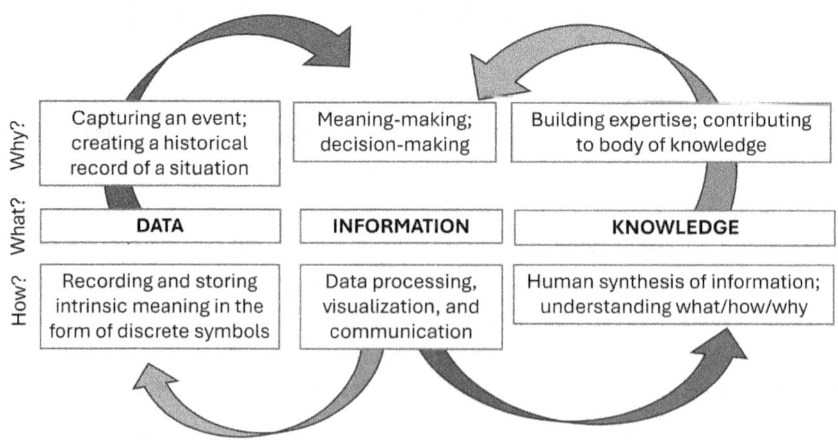

FIGURE 2.1. The relationship between data, information, and knowledge (adopted from Liew, 2007).

TABLE 2.1. ***Definition of the Researcher***

TRADITIONAL DEFINITION	RENEWED DEFINITION
A researcher is someone who conducts systematic investigations.	A researcher is a generator of knowledge and a steward of data. They gather data ethically without exploiting, make claims with integrity, and take care of data before passing them on to future stewards.

this recognition, the procedures needed for responsible data management are often put on the back burner. Without an idea or plan for the use of or how they will be stored and shared, data practices can be likened to hoarding various objects. This is a dangerous situation for research integrity and a recipe for wasting resources. Hence, in Table 2.1 we present a renewed definition, one that situates data stewardship as a core element of research. The explicit recognition of this responsibility is important to communicate with new researchers.

Through an emphasis on data stewardship in the definition of a researcher, our aim is to grant higher significance to the care, reusability, and preservation of data. According to Wilkinson et al. (2016), data management and stewardship are the "long-term care" of digital assets, so they can be retrieved and reused in the future. Quality data gathered by a primary generation of researchers can be continuously reused and expanded upon by future generations of researchers, and as such, it is the duty of the current generation to ensure the organization, preservation, and management of that data until it is passed on to future researchers.

2.3 WHAT IS DATA STEWARDSHIP?

Data stewardship is the care and management of data, information, and knowledge throughout the life cycle of a research project. Such stewardship ensures the researcher's ability to generalize data-based insights into knowledge. In addition, the ability to apply disciplinary first principles to data analysis practices is essential, along with data management practices such as the ethical collection and proper archival of data. As stated by Wilkinson et al. (2016), "data stewardship includes the notion of 'long-term care' of valuable digital assets, with the goal that they should be discovered and re-used for downstream investigations."

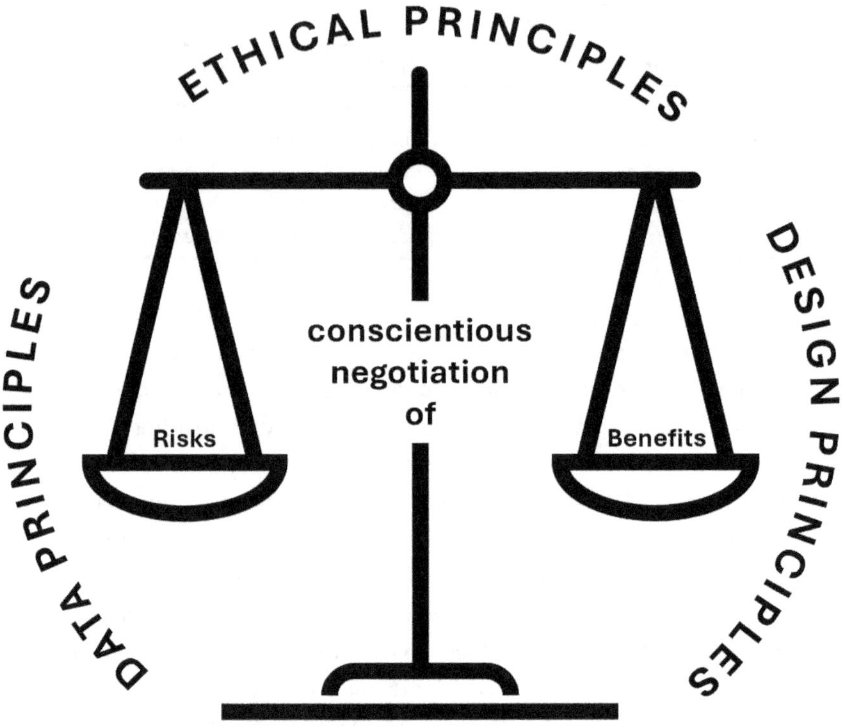

FIGURE 2.2. Conscientious negotiation with data, ethics, and design reasoning.

Caring for and managing data also require making trade-off decisions (e.g., prioritizing accuracy over size of data) and refining decisions through futures thinking (e.g., anticipating unintended consequences in the future).

Research fields differ greatly in subject and priority, and they each place emphasis on different areas of data stewardship. In the medical field, data stewardship has a strong focus on respecting the legitimate rights and interests of stakeholders and the responsibility to maintain a fair and trusted data ecosystem (Wendelborn et al., 2023). While these aspects are also important in engineering and science, there is also a notable emphasis on reusability and management of data. To capture these nuanced variations, we conceptualize data stewardship as a form of reasoning (Purzer et al., 2024). As represented in Figure 2.2, effective stewardship requires conscientiously negotiating risks and benefits. And at a broad level this negotiation takes place by concurrently reasoning about data, ethics, and design.

2.4 THEORETICAL FOUNDATIONS OF DATA STEWARDSHIP

To effectively achieve long-term care of data, data stewards need fluent reasoning and deliberate practice. The framework we developed is designed to help manage the cognitive challenges associated with data stewardship. As depicted in Figure 2.3, data stewardship represents the intentional integration of data principles, ethical principles, and design reasoning in the everyday practices of researchers. The framework also informed the development of a toolkit designed to scaffold the deliberate practice of data stewardship.

DATA PRINCIPLES

Data principles are composed of four aspects: findability, accessibility, interoperability, and reusability (FAIR). The FAIR principles represent the technical responsibilities of data stewards (García-Closas et al., 2023). It is essential to responsibly curate, use, and share data, as these efforts contribute to the historical record of events. Data analysis, then, serves as a deliberate act of interpreting events by applying different analytical lenses. Furthermore, data life cycle management is the responsibility of the researcher, who must ensure accuracy, legality, fairness, privacy, and security of data and the legitimacy of information derived from these data (Gundersen, 2017).

- *Findable* means that data should be easily discoverable. Metadata are key to finding data.
- *Accessibility* refers to being able to retrieve data once it is found. This can include the use of data formats that allow broad access.
- *Interoperability* refers to compatibility with other databases. Common data structures and vocabulary help with interoperability.
- *Reusability* refers to further uses of data and the replicability of research. Clear descriptions, strong documentation, and clear usage licenses help with the reusability of data.

ETHICAL PRINCIPLES

Data management principles and ethics are inherently connected. The separation of this binding has led to systematic discrimination against certain content and groups, with severe impacts on society. While data may look like discrete symbols, the meaning created from the analysis of these data lead to claims and decisions that impact people such as credit scores, health care services, laws, and regulations (Obermeyer et al., 2019). Reflexive principlism, developed by Beever and Brightman (2016), is a framework that guides ethical reasoning by internalizing a reflective and iterative process. It uses a set of four common ethical principles (Beauchamp and Childress, 2009) that need to be contextualized, justified, and balanced:

- *Beneficence:* increasing benefits while preventing harms
- *Non-maleficence:* avoiding and not creating harm
- *Autonomy:* respecting autonomous decisions of research participants
- *Justice:* distributing benefits, risks, and costs fairly

In reflexive principlism, the practice of reflection is essential for ethical decision-making. Beneficence guides researchers to prioritize the prevention of harm while enhancing societal benefits. Non-maleficence refers to avoiding actions that could cause harm and engaging in reflection to foresee potential future harms. Autonomy emphasizes respecting individuals' right to make autonomous decisions with respect to how data created by them are used. Justice stresses the importance of fairness in distributing benefits and risks as well as costs. While the importance of ethical reasoning in research is well established, ethics education is frequently treated as distinct from disciplinary technical content. Our framework closes this divide by positioning design reasoning to interweave data and ethical principles.

DESIGN PRINCIPLES

Conscientious decisions require fluency in reasoning across different forms of knowledge (Purzer & Arigye, 2024; Quintana-Cifuentes & Purzer, 2022). The design reasoning framework offers four quadrants that help activate conscientious negotiation and bridge data and ethical principles:

- *Experiential reasoning* refers to generalizations that a researcher arrives from experiential learning or past lived experiences. With such reasoning, researchers can tap into their practical experiences and form insights from observations, which can result in novel ways of handling data.

- *First-principles reasoning* refers to key disciplinary, data, or ethical principles that guide decisions around data. In this quadrant, researchers also rely on theories within their discipline to guide their efforts.
- *Trade-offs reasoning* refers to the factors that are taken into account while negotiating the risks and benefits of decisions about data. Researchers weigh conflicting requirements and the pros and cons of different data and design choices.
- *Futures reasoning* refers to the ability to look ahead and imagine decisions in new contexts and judge future implications or societal impact of data management decisions. Such reasoning involves imagining implications of decisions in new situations and future contexts.

When connected with data and ethical principles, these four quadrants of reasoning facilitate agency and conscientious decision-making. This integration

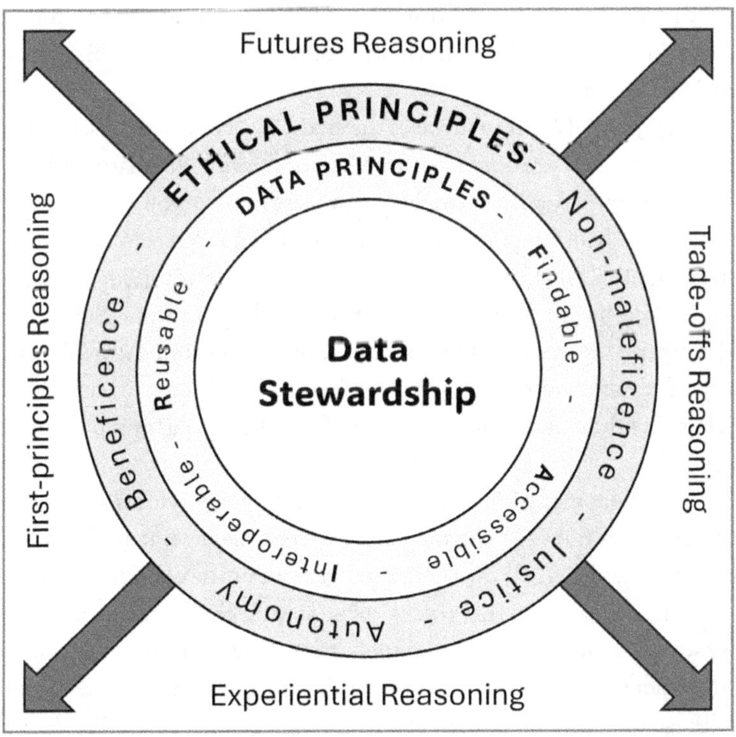

FIGURE 2.3. Data stewardship framework.

is what ultimately establishes the data stewardship framework (see Figure 2.3), connecting how we reason, why we care, and what we act on.

2.5 RESPONSIBILITIES OF A DATA STEWARD

For researchers around the world, access to legitimate data is becoming harder to source. Typically, universities and companies employ specialized experts as data stewards (Rousi et al., 2024). However, as we argued earlier, everyone involved in the use of data should also be responsible for the care and management of data. Hence, from undergraduate researchers to principal investigators of research, all involved have the responsibility of data stewardship.

Data gathering is a process that involves time-intensive steps such as recruiting participants, conducting experiments, and interviewing participants. Web scraping methods used with AI can speed up data collection and help achieve data sizes for effective analysis. Software designed to generate stand-in data points are commonly used as information when real-world data are too expensive to obtain. However, these shortcuts create risks in terms of bias, relevance, and participant consent.

Another potential concern involves the analysis of collected data. It is through analysis that the data are converted into information. While large datasets can be processed using AI analysis tools, the process needs to be supervised by human researchers. Most importantly, the creation of knowledge must be led by humans (see Figure 2.1). The researcher's role as the generator of knowledge and a steward of data has further significance in an AI-rich future. Such concerns should explicitly be discussed in research teams. To ensure the quality and legitimacy of research, researchers must gather data ethically without exploiting, make claims with integrity, and take care of data before passing them on to future stewards.

While there are many broadly established data analysis tools and methods, a similar trend does not exist when it comes to data management. As a result, research teams establish their own data management protocols. Creating protocols from scratch can lead to higher risk of encountering challenges in data management practices. Our hope is that the toolkit we present in this book will be useful to the readers while coordinating their data management efforts across their teams. These tools can be refined and adapted to the users' specific data needs. While the toolkit offers practical guidance, the data stewardship framework provides a conceptual foundation for reasoning about ethical management of data. We hope that mentors can support their mentees with this integrated reasoning framework in their formal and informal conversations and coaching sessions. Chapters 3, 4, and 5 include a range of methods and tools that aim to make data stewardship practices accessible to all who work with data.

3

DATA STEWARDSHIP GUIDE FOR RESEARCH MENTORS

3.1 SUMMARY

This chapter provides a guide for mentors to assess and develop data stewardship skills among students using practical tools such as a rubric and reflection questions. The combination of rubric-guided assessment and reflective questioning promotes not only technical skills but also ethical and strategic reasoning necessary for responsible research data management. An example from a student research team working on classifying deer bone fragments will be provided to illustrate practical application of the tools.

3.2 PRACTICAL TOOLS FOR ASSESSMENT AND MENTORING

An effective practice and operationalization of the data stewardship framework is the data steward rubric presented in Table 3.1. This tool can be used by research mentors to guide and assess key elements of data stewardship reasoning. Although the rubric was developed initially to assess data steward competencies in a poster session setting, it can be used in both formal and informal conversations to articulate specific abilities and activities associated with each of the competencies.

In addition to the rubric, the reflection questions provided in Table 3.2 can be used to help students develop reasoning skills and build confidence to negotiate the complex and interconnected aspects of data stewardship.

TABLE 3.1. ***Data Steward Rubric***

	MASTERY EXPECTATIONS	EVALUATION
Experiential Reasoning	❑ Explain the data analysis and visualization methods (charts and tables) used ❑ Explain the nature of the data that led to the visualizations presented ❑ Provide examples of applying past lived experience and application of prior knowledge ❑ Present examples of insights and strategies learned experientially; there is evidence of new learning	❑ Proficient (4 met) ❑ Advancing (3 met) ❑ Developing (2 met) ❑ Emerging (1 met) ❑ None
First-Principles Reasoning	❑ Articulate key technical principles that informed data management for this applied research project (e.g., the data collection, preservation) ❑ Articulate key ethical principles that informed data stewardship for this applied research project ❑ Present evidence of planning to ensure data are findable, accessible, interoperable, or reproducible ❑ Provide details of how first principles impacted project design decisions (required data sample size for AI model training, proper methods for data analysis, metadata creation, privacy, data storage, sharing, etc.)	❑ Proficient (4 met) ❑ Advancing (3 met) ❑ Developing (2 met) ❑ Emerging (1 met) ❑ None
Trade-Off Reasoning	❑ Recognize existence of a range of trade-offs and challenges as part of data life cycle (data collection and management, consent, IP, ownership, storage) ❑ Articulate competing issues, where improving one aspect hurts another one (accuracy vs. cost, time vs. privacy, etc.) ❑ Provide details of a specific trade-off decision made during data collection, management, analysis or preservation ❑ Articulate how ethical principles are balanced within decision-making	❑ Proficient (4 met) ❑ Advancing (3 met) ❑ Developing (2 met) ❑ Emerging (1 met) ❑ None

(continued)

Futures Reasoning	❑ Describe how data management decisions position a dataset for long-term preservation and sustainability ❑ Share lessons learned about data practices to carry forward ❑ Predict potential misuses and applications of data or knowledge generated that could result in future harm ❑ Articulate the broader, societal benefits of the data and knowledge generated with the data	❑ Proficient (4 met) ❑ Advancing (3 met) ❑ Developing (2 met) ❑ Emerging (1 met) ❑ None

TABLE 3.2. ***Data Stewardship Sample Reflection Questions***

DATA STEWARDSHIP REFLECTION QUESTIONS	
Experiential Reasoning	Could you provide me a high-level overview of your data workflow? Can you share how prior experiences—whether from coursework, previous projects, or lab work—influenced your approach to handling data in this research? What are examples of insights and strategies that you learned through this experience?
First-Principles Reasoning	What key disciplinary principles guided your decisions around data collection and management, and how did they shape your project outcomes? What were the specific meanings of ethical principles (e.g., beneficence, nonmaleficence, autonomy, and justice) related to data for this project? Now that you completed this project, what principles or takeaways do you have about conducting research?
Trade-Off Reasoning	What challenges have you faced, especially when weighing trade-offs with respect to data collection, analysis, or storage? How did you balance the ethical aspects and principles related to the project data? What specific practices are important for ensuring ethical research? How did you address issues like privacy, data sharing, or intellectual property?
Futures Reasoning	Looking ahead, how do you see your data stewardship practices impacting the future use of your data? Can you describe how your data are organized, documented, and stored from collection through analysis and final outputs so that others could find, access, and reproduce your work in the future? What are potential societal impacts of your research?

3.3 A DEEPER LOOK INTO THE PROCESS OF DEVELOPING COMPETENCIES IN DATA STEWARDSHIP

This section presents a detailed example from an undergraduate research team and demonstrates not only the impact of the use of the data stewardship toolkit on the students' project but also how the use of reflections before and after using the toolkit impacted the students' data stewardship understanding.

The dataset for the project included fragments of radius and tibia bones from deer, and the team's task was to develop a system for classifying each bone fragment appropriately. The team explored different image processing and machine learning approaches for this task, with their final product able to classify the two different types of deer bones with 92% accuracy. The team presented their work at the Purdue Undergraduate Research Symposium in spring 2025, where they were selected as one of the inaugural data stewardship awardees. (For details, refer to https://www.purdue.edu/undergrad-research/conferences/spring/archive/past-winners.php.)

In addition, the students on the project published a research summary in the *Journal of Purdue Undergraduate Research* (*JPUR*) (Jain et al., 2025). Two students from the team completed reflections before and after completing several tools from the data management toolkit provided in Chapter 5. The learning that they express in their reflections, as well as the notable changes in their use and understanding of the data stewardship framework, provides insights that can help guide the mentors of undergraduate researchers.

3.3.1 INITIAL REFLECTION

At the end of the project and prior to completing the toolkit, the two students were asked to respond to a subset of the questions from Table 3.2. In response to the question "Could you provide me a high-level overview of your data workflow?" the students described their data workflow, as shown in Figure 3.1.

Their reflection provided more information and insight into their understanding of and approach to data management. (Note that quotations from students in this chapter have been lightly edited to align with other grammatical choices made in the book.)

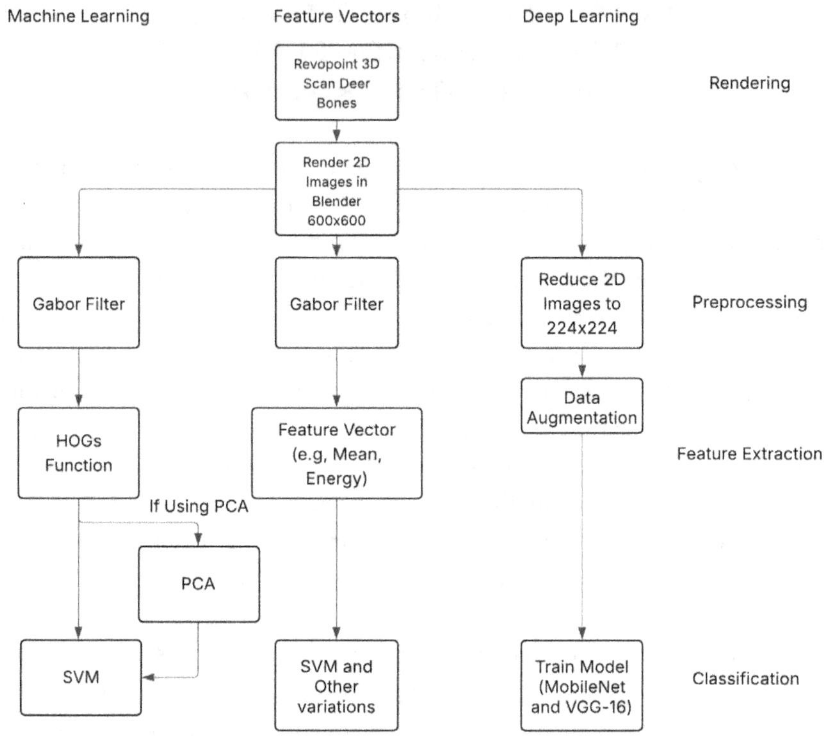

FIGURE 3.1. Initial Deer Bone project workflow.

Note: HOG: histogram of oriented gradients. *PCA*: principal component analysis. *SVM*: support vector machines. MobileNet and VGG-18 are convolutional neural network architectures.

When we converted our deer bone samples into 3D models, we used the Revopoint 3D scanner, which we used to capture 1,200 frames of each bone as it rotated on a turntable and create a dense point cloud. We then used the Revoscan 5 software to turn this dense point cloud into a 3D model of the bone. Since the bone had to sit on the aforementioned turntable, we would scan the top and bottom of each sample separately, then line the models up by matching features on each side of the bones. The whole process took about 20–30 minutes per bone.

The fragments we were given consisted of two major varieties: large and small fragments. Large fragments were larger and had one of the joints of the bone intact. Small fragments were shorter and taken from the center of the bone, often consisting of little more than a shard of bone ~1″ long. This caused some unique issues with our meshes as large fragments were much easier to scan than small fragments since they had a larger surface area to pick up on, whereas the scanner would sometimes have difficulty picking up small fragments and catch part of the surroundings instead. If the scanner couldn't pick up a certain spot on the fragment during the scanning process it would result in a hole in the mesh which needed to be patched, which caused a small loss of data in the model making process.

We then loaded this mesh into blender and used blender's build in coding interface along with the Bpy module in Python to automatically render 600×600 PNGs of the deer bone from 100 different angles. Due to the way we scanned in the bones and recreated them from a point cloud, we did not maintain the bones' color. Despite the bone meshes themselves not having any color, the way we rendered the images in blended did not take this into account so it created color channels for the images; however, we rectified this by converting these images back to gray scale in Python. In total we scanned 55 bones that we separated into three testing and training datasets. We chose these datasets in order to combat data imbalance as we had many more tibias than radii. Additionally, these datasets consisted of an even mix of large and small fragments. We created a 4 tibia and 4 radii dataset with 2 training bones—one of each type—which was simple to implement and easy to compute. Then we created a medium benchmark for all our methods, which consisted of the 9 tibia, 9 radii, and 4 equally split training bones. Lastly, we used an imbalanced dataset consisting of all our available data, which was severely imbalanced towards tibia. It consisted of 24 tibia and 8 radii for training, and 20 tibia and 3 radii for testing. Depending on the method of machine learning or deep learning, these images were handled differently.

For SVM [support vector machine] + HOGs [histograms of oriented gradients], the images are passed through a Gabor filter and then compressed into an array of features by the HOGs function. These HOG features are then compressed again by PCA (principal component analysis) into 100 dimensions per bone. This essentially created a list of compressed features that corresponded to a single bone, which would then be added to a list that comprised all the bones in the training data, which would finally form the basis for our machine learning methods to classify the testing data with. This testing data was also compressed using the PCA. The compressed features allowed the SVM to not overload the computer when running as due to the way we processed the HOG function as it would create an excess amount of data that forced the SVM to fit a line to several thousand data points. For the SVMs without the PCA, we had to drastically simplify the HOGs function by decreasing the size of the cells we used to sort the gradients of the image because of the potential memory overflow. The SVMs used the rbf (radial basis function) kernel for data processing.

The deep learning methods were much simpler in processing the 2D images. We used two different CNNs [convolutional neural networks] to compare performance to, each with their drawbacks and strength. Both used the raw gray-scaled 2D images of the bones in their network. Each network input used a RGB 3-channel image 224×224×3 input, so to adapt our data to the network easily we passed the images through the red channel of each of the networks. In addition, the images were sized down to 224×224, and to account for class imbalance in the datasets, the radius images were weighted 3 for every 1 tibia image. Data augmentation was also added to each network, including contrast and brightness adjustments, random flipping, and rotation.

Lastly, we used feature vectors as a data processing method to further condense the data down for our machine learning methods. These feature vectors calculated a set of simple features from each bone after they were modified by the Gabor filters. Because of the simplification, the data could be used with other less well-known methods such as XGBoost, KNN, Logistic Regression and others.

Each of these methods were tested with the three training and testing sets. The accuracy results were then visualized by a bar graph for the machine learning result and a line graph for the deep learning results.

When asked to "tell about the data journey from the data collection, through the data analysis up to the data output (probably visualization)," the students responded:

Our team had first received a set of 16 deer bones from the Department of Anthropology to use in our machine learning algorithms and deep learning algorithms. One of the original bones we received was a humerus instead of the typical radii and tibia we were using, so we excluded it. This set of data included

3D scans of deer bones using the Revopoint scanner that we would eventually use to scan the rest of the deer bones. Originally, we did not have a workflow for the data as we had not yet decided on a clear plan for processing our data. The main challenge using the data was converting the 3D models of deer bones to a 2D image with enough detail to use for both machine learning and deep learning. Eventually, we decided on a process which involved taking photos at multiple angles of the bone to feed into the algorithms. This allowed us to map the shape of the bone in 2D and increased the amount of data we would put into the data-intensive algorithms.

As we scanned more bones alongside students in the Department of Anthropology, we received a secondary dataset a few months into the project which consisted of the aforementioned 55 bones. This addition to our dataset did cause an issue in our file naming scheme since the later bones were identified using a 4-digit number along with the date they were collected and the type of bone it was, whereas the previous dataset did not have that information so we identified them based upon their storage location in our files. While this did work for our project as it was small, as it grows, this could cause issues finding the old dataset.

Once we had the extended dataset, we analyzed the data we had gathered and noticed a large imbalance of classes. The radius was 1:4 to the tibia (11 radius and 44 tibia), which created a large concern about the performance of the models as the likelihood of overfitting especially for the machine learning algorithms would dramatically increase. This imbalance happened due to the availability of bones within the Department of Anthropology—radii were far less plentiful within the department. As such, the challenge of overfitting within our models became seriously problematic. To solve this overarching challenge, we split the datasets into the groups mentioned in previous segments to avoid the imbalance as best we could, especially in testing as overfitting would give false results in our accuracy tests. In addition, for our deep learning methods we weighed the radius 3 times more than the tibia when there was an imbalance.

We chose to use two separate types of charts to display our result because of the difference between machine and deep learning. Machine learning uses the dataset a single time to create points with which to compare the testing data to, whereas deep learning uses the same training data repeatedly in epochs to get a better performance out of the model. This means that the result from the

> machine learning model was a single number, whereas the results from deep learning would change based upon how many times this process of training the model went on. Based on this, we used a line graph to chart the accuracy of our deep learning methods as the training went on.

Although Figure 3.1 primarily focuses on the data analysis activities, the students' reflections provided more insight into their data management activities and articulated the different approaches they used, how they partitioned the dataset for testing and training, and challenges they encountered because the dataset was limited and imbalanced. They noted some challenges such as file naming issues, initial confusion on bone type labels used in code, and inconsistent testing datasets that made comparison among the different methods difficult, which they worked to standardize. Through these discussions, they demonstrate first-principles, trade-offs, and data reasoning skills.

In response to the question "Based on the visualization (or key highlight in the poster), can you provide me with the key take away of your project/results?" the students reflected as follows:

> The value of data outweighs the quantity of data. In other words, data should always be made to be as valuable as possible rather than becoming stretched and expanded, and effort must be made to minimize any possible data loss due to the processing of this data. In our project, the data processing method with the best results was the feature vectors, which condensed the data into informationally dense features. When compared to HOG + SVM, the feature vectors surprisedly performed well in every metric because the vectors managed to condense the data better than the PCA. In the pursuit of simplifying our images into recognizable data to train our machine learning methods with, even without PCA the HOG method was removing small amounts of data by normalizing the histogram results across multiple blocks. The PCA, as seen by the final results, demonstrably worsened the accuracy of the SVM as information was lost in the preprocessing stage. This indicates the value of trying to extract as much information as possible from the given dataset and being careful not to lose data in the pursuit of computational efficiency.

One key takeaway the students identified in this reflection was the importance of preserving as much information through data processing as possible to achieve better results versus trying to achieve computational efficiency. This demonstrated experiential and trade-offs reasoning.

The students' reflections also identified ethical issues and practices they encountered related to supervision while scanning bones and the practice of storing data on hard drive versus online. Students identified the need to disclose the performance of the different approaches they used in their analyses for transparency as an ethical issue.

3.3.2 IMPACT OF USE OF TOOLKIT

After completing the initial reflection, the students were given the opportunity to rework and improve the data workflow of the project by utilizing a variety of tools from Chapter 5 of this book. The toolkit enabled the students to develop data storage and management plans and identify issues that had not been addressed when working on the project.

In response to the prompt question "Could you provide me a high-level overview of your data workflow?" the students indicated that the first toolkit they completed was the Data Management Planning Discussion Guide (see Chapter 5, Section 5.2), which prompted them to explore more comprehensive aspects of data management and identify gaps in their understanding of the data workflow. As a result, after completing the toolkit the students made several improvements related to their data management practices.

The first improvement was to migrate their dataset storage from a personal cloud-based drive system, which had storage limitations, to a content management system that is supported by the university and, thus, has more storage capability. Because of the space limitations, the students noted that they needed to compress their data before uploading, which did not allow them to view files without downloading the entire dataset. Although the students identified this as an additional backup to their personal computers and lab server is an improvement, this created a new gap, a process to ensure consistency across these dataset storage sites.

The second improvement was related to file naming and organization. The students used the File Naming and Organization toolkit (see Chapter 5, Section 5.6) to organize the data and changed the naming convention of the image file. Their completed toolkit is shown in Table 3.3. The naming convention used for the original dataset was Box#_TimeGenerated_BoneID. The convention was changed for the expanded dataset to YYYYMMDD_Bone###_BoneTYPE_DataFormat

TABLE 3.3. *File Naming and Organization Toolkit for Deer Bone Project*

DATA CONTENT	DATA DEVELOPMENT	EASE OF ACCESS	OTHER
What categories of data will be generated in this project? Code, [missing info].	*Will your data go through multiple stages of development?* Yes, our image data will be processed using a variety of techniques to improve its use in identification models.	*What data categories will be accessed most frequently?* The code, image data, and final report will be accessed the most by future researchers.	*How will your team ensure that data organization is upheld?* We will have a section in the metadata document to enforce file structure and naming schemes.
How will each type of data be documented, standardized, and stored? [No response.]	*Will your data be processed?* Yes, the data starts out as 3D models and is rendered into 2D images before being processed into results.	*How can you organize your data so researchers can access data quickly?* We will use a file naming scheme that makes it easy to identify the file and a folder system to separate our data from our results.	*When will your team organize the database?* The basic structure of our organization was outlined at the start of the semester, after which we filled in data as it came to it.
What formats of data will be used (Excel, MySQL, jpg, etc.)? PowerPoint, Google Docs, Python, 3D models, and PNG images.	*Will you need an organization section for raw data?* Yes. We will use a folder system to contain our data.	*In what database will the data be stored?* Originally it was in the basement hard drive, remote access server, and [private] drive. Currently we have improved upon the ease of access by using [university drive].	*How will you preserve or "back up" data in preparation for accidental data loss?* Bone data is backed up on a hard drive with the Department of Anthropology, and on a server with [research] Laboratory, and on the cloud via [university drive].

(continued)

What metadata will be included with your data? We will include a full writeup of our data in accordance with DCMI standards using the Cornell template.	*Will you need temporary storage for developing data?* Yes. The unprocessed bone scans are stored on a hard drive in the Department of Anthropology.	*Who will have access to the data?* Students and professors with the Department of Anthropology and [research lab].	*Will you require storage for other files or data documentation?* Yes. We originally used [team file drive] for our PowerPoints and [personal drive] for our final report; however, we have switched to [university drive] for all of the above to keep things straight.

to include metadata associated with the image. This change addressed issues students encountered while working on the project. Not including the bone type in the file name originally led to confusion and errors when processing the data.

The toolkit also guided them to create a new file organization system, as shown in Table 3.4. This new system allows for easier access to all the team's materials, including weekly presentations and project documentation in addition to the data, using two data storage and collaboration spaces (document cloud storage and code version control system). The project code was stored in a version control software system that was hosted on the research lab's servers.

The third improvement the students made was the result of recognizing that they had not organized the results from their analyses in a way that enabled comparisons across different trials. Although the students would present their results via regular slide presentations, they did not have a centralized location to store the data so that they could make appropriate comparisons and interpretations of

TABLE 3.4. ***Deer Bone Project File Organization System After Completing Toolkit***

CLOUD DOCUMENT SYSTEM		CODE VERSION CONTROL SYSTEM
DATA	**TEAM PRESENTATIONS**	**SOFTWARE**
Object files	Archived PowerPoints	Models
Rendered images	Final report	Data
	Research expo poster	Processing methods

the results across the different trials and changes to the development of the analytical approach. It is interesting to note that the students had previously encountered challenges to their ability to present and compare their results when writing the research highlight for *JPUR*, but they did not recognize at the time that this was a data management issue or have strategies for addressing it. And although they demonstrated awareness of this issue after completing the toolkits, strategies for managing the results are missing from their revised project workflow, as well as from other data management toolkit artifacts. These gaps illustrate challenges to incorporate data management practices throughout the project and including results.

The fourth improvement needed became clear when students reflected on how the concepts of data description and data dictionaries helped them improve the legibility of their dataset. They used Cornell University's README template (Cornell Data Services, n.d.) to document essential information about the dataset at the project level, which included contacts for the project, descriptions of the data, methods used for scanning the bones into 3D models and the processing to convert it to 2D images, and defining the file naming conventions. The completed README file was saved as a .txt file so it could be read by any computer. In addition, they completed a data dictionary as shown in Table 3.5. However, this draft of their data dictionary focused on defining the names of the data analysis approaches, such as PCA and SVM, versus describing the metadata, variables, and types of data used and generated in the project, as illustrated in Chapter 5, Section 5.3.1. This gap highlights an area where undergraduate researchers may require additional guidance from mentors to effectively implement these data management practices.

The fifth improvement came about when, from completing the toolkit, the students recognized the lack of clarity on data rights. Although they had utilized resources in the Department of Anthropology under the guidance of the faculty mentors to create the dataset with the expectation that the data would be used by future teams and researchers, the reuse permissions and copyrights were not explicitly discussed. It is recommended that mentors provide clear guidance and expectations regarding data use at the outset of the project, which can be facilitated through completion of the toolkit.

Beyond the changes and learning related to data workflow and management, the reflections provided insight on learning related to the data stewardship framework. When the students were asked to reflect on first principles after using the toolkit, the key disciplinary principles related to image processing (i.e., the need to retain data and the impact of the limited and unbalanced dataset on their results) did not change. However, they identified new principles related to

TABLE 3.5. *Deer Bone Data Dictionary*

VARIABLE	DEFINITION IN THE PROJECT	WHERE DATA ARE USED THROUGHOUT THE PROJECT
Radius/radii	The radii proximal or upper foreleg bone of the deer. Also denoted in code as 0.	Everywhere.
Tibia/tibiae	The tibia distal or upper back bone of the deer. Also denoted in code as 1.	Everywhere.
Accuracy (%)	% of correct items/total # of bones in testing dataset.	Weekly reports, final presentation, and report.
PCA	Principal component analysis. A method of reducing dimensionality in data.	Weekly reports, final presentation, and report.
SVM	Support vector machine. A bone classification method we tested.	Weekly reports, final presentation, and report.
Gabors	Processing tool to enhance edges in images.	Weekly reports, final presentation, and report.
HOG	Histogram of oriented gradients. An image processing tool used to better define edges.	Weekly reports, final presentation, and report.
Cells	In HOG, the size of the segments the image is split into to process.	Weekly reports, final presentation, and report.
Blocks	In HOG, the size of the overlapping cells being normalized with each other.	Weekly reports, final presentation, and report.
Conv.	Convolution. A step in neural networks where a filter is applied to an image.	Weekly reports, final presentation, and report.
Maxpool	A downsampling method that takes the largest value over an area.	Weekly reports, final presentation and report.
Softmax	The output function of a neural network where data is converted to binary results.	Weekly reports, final presentation and report.

data management, which included file organization, better data control, and transparency regarding data sourcing and acquisition. This outcome pointed to the benefits of the toolkit in helping students implement these principles in their work.

In addition, the completion of the checklists helped broaden the students' understanding of their ethical responsibilities related to data management. Prior to completing the toolkit, the students identified ethical issues related to data collection, which included responsibilities and oversight needed when scanning the bones, the need to keep the bones in the lab, and the expectation to store the data locally. They also described the considerations the system should use to classify human bones in the future. In their post-reflections they also included their recognition of their obligation to ensure that the data are readable, accessible, and transparent to enable researchers to use the data in the future. One explicit aspect they mentioned related to transparency was the need to include the metrics of the model itself and its testing. In addition to the ethical implications of adhering to these principles, the students noted that practicing the principles also facilitated onboarding of future researchers and comparison of methods.

In the post-reflection, the students responded to the prompt "What ethical concerns did you encounter in managing your data, and how did you address issues like privacy, data sharing, or intellectual property?" as follows:

> Finally, while our project revolved around deer bones as a proof of concept, ideally in the future this project would lead to analysis of human bones as well. When conflicts lead to loss of life overseas, human remains can take years to return to the U.S. This identification of human remains is a difficult and arduous task, which has led to a backlog of cases that can be resolved by automatic classification models such as these. However, this opens up numerous ethical concerns for the collection, storage, and accuracy of the data. Consent would have to be obtained from the living in order to undertake any form of experimentation with human remains; storage and management would need to be extremely strict so that all remains can be returned to their next of kin once the research has concluded. Even after collecting this data, care needs to be taken to avoid accidental mix-ups which would lead to families being given false information. While these concerns were minimized in our project by using deer bones, if this project were to eventually extend to human remains, these ethical concerns would need to be revisited.

Their post-reflection expanded the considerations needed if the system were to be used to identify human bones. Specifically, they identified the need to obtain consent from next of kin and for the utmost attention to accurate tracking of remains and reporting back to families. This reflects a recognition of broader, societal benefits of the research as well as ethical implications of such research. There are also opportunities to promote futures reasoning in terms of possible, probable, and preferred applications of data or knowledge generated with the data.

3.4. CONCLUSION

The example provided in this chapter, with a deeper examination of a team's development as data stewards, highlights how completion of the data management toolkit and reflections can help students improve their data management practices and data stewardship learning. It also highlights the need for mentors of undergraduate researchers to explore data management practices in addition to providing feedback on the data analysis aspects of the project. Many aspects of data management are hidden and not traditionally the focus of research mentorship and feedback, but as shown in this chapter, data stewardship is critical for effective implementation, sustainability, and reproducibility of research projects.

4

DATA STEWARDSHIP IN PRACTICE

4.1 SUMMARY

This chapter presents three artificial intelligence (AI)-focused undergraduate research cases illustrating the application of a data stewardship framework through poster sessions. The analysis highlights students' strengths in reasoning about data management, trade-offs, and ethical considerations, reinforcing the value of experiential learning in undergraduate research.

4.2 ASSESSMENT OF DATA STEWARDSHIP

POSTERS FOR FORMAL ASSESSMENT

Researchers often engage in formal presentations. For undergraduate research, the format is typically in the form of a poster session. These poster sessions allow focused conversations and feedback from experts. A poster session allows the researcher to present their work and the expert to ask questions and seek clarification. During a poster session, the presenters stand by their poster and provide verbal descriptions of the work depicted. Although sizes differ, a typical poster is 36 by 48 inches and includes elements such as title, abstract, methods, and results. When there is formal judging, judges utilize a rubric that focuses on the elements of the research and the presentation.

With an effort to illustrate the operationalization of the data stewardship framework, this chapter presents examples of actual student research posters presented at an undergraduate research symposium held at Purdue

University. These posters were self-nominated for consideration for the data stewardship award. Each poster example presented includes three components: a title and abstract, answers to three targeted prompts, and a research poster. This chapter highlights the strengths of three artificial intelligence (AI)-focused posters with respect to a few key aspects of data stewardship. These are the poster topics:

- Poster Example 1: Emergency drone delivery of Narcan
- Poster Example 2: Bibliometric analysis of cancer research trends
- Poster Example 3: Statistical analysis of fantasy basketball

VALUE OF DISCOURSE BEYOND THE POSTERS

During the undergraduate research symposium, the authors talked with undergraduate researchers and asked questions that elicited reasoning for the data stewardship award. Although constrained by time, these conversations were rich and yielded important insights into the students. In particular, it was evident that student researchers who actively participated in large data creation were able to articulate more integrated understandings of data management and technical and ethical principles. They demonstrated more fluency across different reasoning aspects of research. They were also able to discuss specific ways that data management decisions impacted their project as a whole and throughout the various aspects of the design. The data stewardship toolkit can further enrich the skill sets of undergraduate researchers, regardless of whether their projects involve large-scale data generation.

4.3 POSTER EXAMPLES AND STRENGTHS IN DATA STEWARDSHIP

See Figure 4.1 for a screenshot of the poster titled Emergency Response Drone for Narcan Delivery.

POSTER EXAMPLE 1
Emergency Response Drone for Narcan Delivery

By Nathan James Arnold and Parth Kailash Dubal, mentored by James Michael Goppert, Nicole Adams, and Nan Kong

Abstract

Since early 2020, when the COVID-19 pandemic began, opioid overdose deaths in the United States have increased significantly, spiking at nearly 108,000 deaths in 2022. And against what experts believed, after lives began to return to normal this number remained high. While Narcan is an effective, easily administered opioid inhibitor, timely intervention is essential, as brain damage from oxygen deprivation can occur within 6 minutes—far shorter than the average emergency medical service (EMS) response time of 17 minutes in some rural Indiana counties. Even the national average sits at 7 minutes; still too slow to help save a life. This project focuses on developing an unmanned aerial vehicle (UAV) system that autonomously delivers Narcan and instructs bystanders to administer the treatment, significantly improving survival outcomes ahead of EMS arrival. With a critical mission window of just 6 minutes, every second is vital. Accounting for dispatch time and a drone flight time of 5 minutes—achieved by designing a UAV capable of reaching speeds of 60 mph and a path-planning algorithm for an optimal flight path—only 30 seconds remain for instructing a bystander on administering Narcan. However, prior human–UAV interaction trials revealed that medically untrained individuals in high-stress situations sometimes required more than 30 seconds. To address this, our research explores design improvements that enhance instructional clarity and response efficiency. By optimizing the way Narcan administration is communicated, we aim to reduce cognitive load and improve usability, ultimately increasing the number of lives saved in opioid overdose emergencies.

Prompts

How did you approach decisions related to data throughout your applied research project? Throughout our research, we conducted human–UAV interaction trials under Purdue's IRB (Institutional Review Board) and analyzed real-world EMS response time data. Our structured experiments were designed to closely replicate the high-stress conditions of opioid overdose emergencies. The data collected from these trials directly informed the design and optimization of our final product, ensuring it meets real-world usability and effectiveness requirements.

What considerations and trade-offs were involved in your decisions? A larger drone could achieve higher speeds and reduce response time, but a smaller, more compact design is crucial for bystander interaction and ease of Narcan administration. Balancing these factors, we focused on optimizing the drone's design to maintain high speed while remaining lightweight and user-friendly. This approach ensures that bystanders can efficiently engage with the drone, ultimately improving the effectiveness of overdose interventions.

If your project has published any data, please provide a URL link. https://engineering.purdue.edu/AAE/Aerogram/2023-2024/articles/28-narcan-delivery-by-drone.

POSTER 1'S STRENGTHS IN DATA STEWARDSHIP

ABSTRACT

- ✓ Articulates the broader, societal benefits of the research (data and knowledge generated).
- ✓ Presents examples of insights and strategies learned experientially (while working with data). There is evidence of new learning.

PROMPTS

- ✓ Articulates competing issues that influence the knowledge generation process, where improving one aspect hurts another one (accuracy vs. cost, time vs. privacy, etc.).
- ✓ Provides details of a specific trade-off decision (associated with data and knowledge generation).

POSTER

- ✓ Explains the nature of the data that led to the visualizations presented.
- ✓ Connects key takeaways of research to the data visualizations presented.
- ✓ Articulates suggestions to address challenges faced.

The strength of this case is trade-offs reasoning and experiential reasoning, with emergent elements in first-principles reasoning and futures reasoning, and opportunities to grow in ethical reasoning.

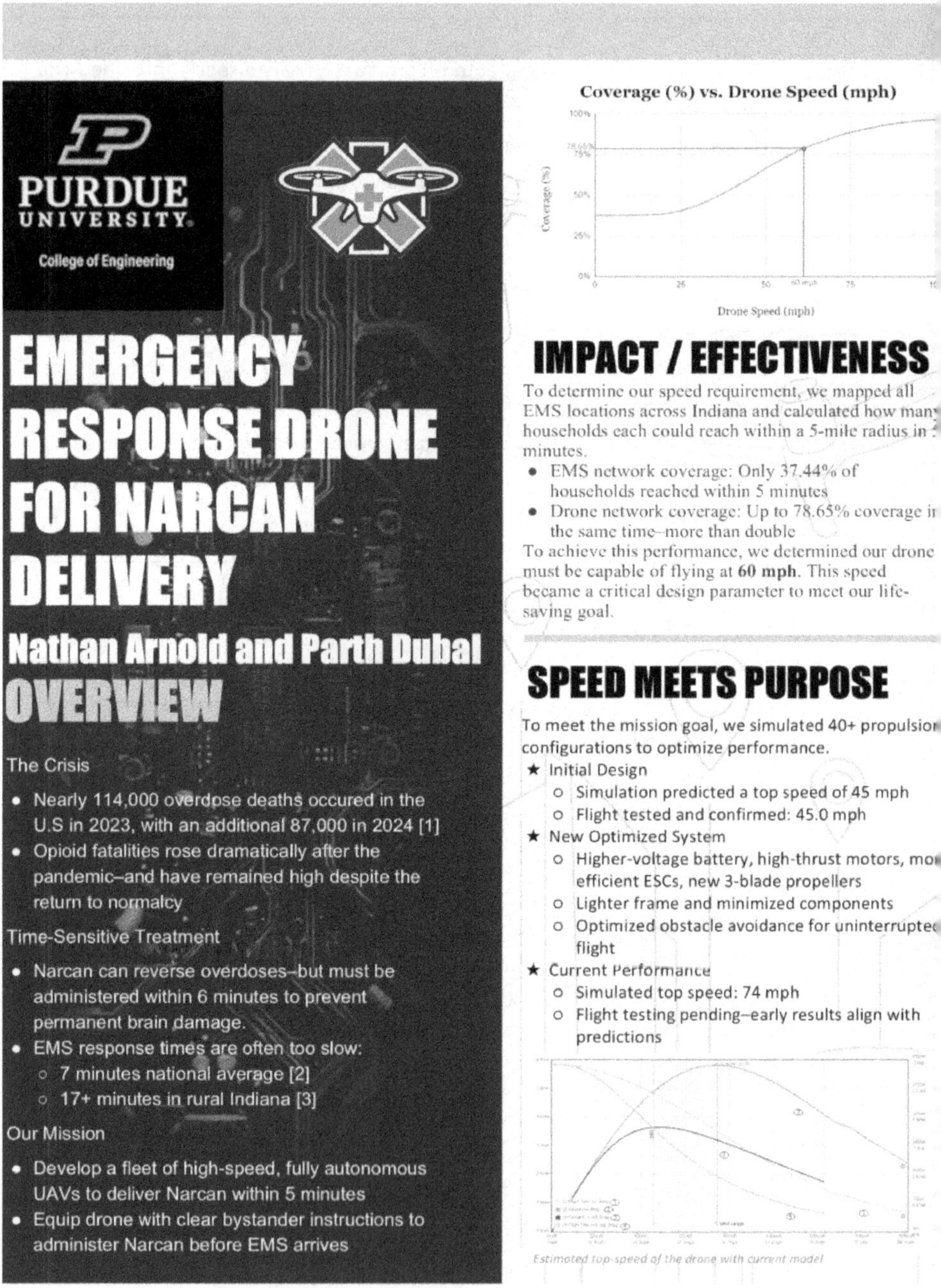

FIGURE 4.1. Poster Example 1. To view a digital version of the above poster, visit https://docs.lib.purdue.edu/stewards/1/.

Participants administering Narcan at first Human-UAV Subject Trial

HUMAN-UAV TEST TRIALS

The Goal: Successfully have participants deliver Narcan to test-dummy in <30 seconds, with 5 minutes already spent on drone flight and 30 seconds for dispatch.
To evaluate real-world usability, we conducted our first human-subject UAV trial at Alexander Field (Purdue Baseball Stadium).
Simulating an Overdose Emergency

- Participants were placed in a realistic, high stress overdose scenario

After "calling 911," a UAV autonomously landed nearby. Participants were then instructed to:

1. Retrieve the Narcan
2. Follow the drone's visual and audio instructions
3. Administer the dose correctly and quickly

RESULTS

- Over 50% of participants were unable to administer Narcan within the required 30-second window
- Instruction clarity and interface design were the primary barriers–especially under the simulated stress
- Participant feedback made it clear:
  - ➢ In its current form, the drone interface risks delaying life-saving action, and its current design is not suited for real emergency use

THREE NEW PROTOTYPES

In response to the initial trial results, we developed three new UAV interface designs aimed at reducing cognitive load and improving clarity during Narcan administration

1. Enhanced Baseline
- Improved screen and speaker for audio and visual
1. Minimal
- Speaker-only interface (AED-Inspired) designed with low complexity and fast use
1. Interactive Model
- Screen + Speaker + Physical Buttons
- Allows user more controllability of instructions

We currently have submitted our proposal to the IRB and are ready for testing our new designs.

CONCLUSION & FUTURE WORK

We are currently conducting trials to determine which of the three redesigned UAV interfaces enables the fastest and most effective Narcan administration. Results will inform any final design adjustments. Looking ahead, we aim to design a 4G-based communication network to support a scalable emergency response system. This will allow:

- 911 dispatchers to launch UAVs as quickly as ambulance
- Real-time identification and dispatch of the nearest available drone, regardless of location

This next phase will bring the project to real-world deployment, starting with select counties in Indiana

References

1. CDC Newsroom, 2025
2. American Medical Association, National Library of Medicine, 2017
3. Indiana Public Law 67 - 2024 Data Report

POSTER EXAMPLE 2

Analyzing Research Trends: Exploring PICR Research Impact, Collaborations, and Expertise With Bibliometric Data

By Colin Roberson, Vineeth Narra, and Lonnie W. Schwartz, mentored by Pete E. Pascuzzi

Abstract

Understanding research output and collaboration patterns is essential for guiding institutional strategy and funding decisions. This study analyzes publishing trends among contributors to the Purdue Institute for Cancer Research (PICR) over recent decades, focusing on citations, interdisciplinary collaboration, data accuracy, and grant allocation. Using R's data analysis and visualization tools, it addresses the challenges of obtaining accurate bibliometric data, which often contains errors from platforms like Web of Science and Scopus. The goal is to enhance understanding of research trends to inform future strategic decisions for institutions. Key deliverables include a methodology for collecting and cleaning bibliometric data, along with advanced analysis and visualization presented in an intuitive web application. This approach can be extended to other domains and institutions to uncover valuable bibliometric trends.

Prompts

How did you approach decisions related to data throughout your applied research project? In this project, we focused on collecting, cleaning, enriching, and analyzing bibliometric data to understand research output and collaboration patterns at PICR. Our goal was to ensure the data was accurate, accessible, and useful in the long term, while also addressing challenges like data completeness, privacy, and stakeholder engagement. We used bibliometric address data to create a more complete dataset by including all contributors affiliated with PICR, minimizing errors from variations in author names. To maintain data accuracy, we used R's data analysis tools to clean issues like mismatched author names and affiliations.

What considerations and trade-offs were involved in your decisions? To enrich the data, we added metrics such as citation counts, coauthor collaborations, and controlled vocabularies like research subfields and grant numbers, which allowed us to apply advanced techniques like network analysis and topic modeling. These methods helped us uncover trends in collaboration and research impact, revealing strengths and opportunities in PICR's research ecosystem. A key part of our approach was creating an interactive dashboard to make the data accessible to PICR stakeholders, NIH (National Institutes of Health), donors, and collaborators. This tool allows users to explore research outputs and collaboration patterns, but we ensured privacy by aggregating and anonymizing sensitive data.

Throughout the project, we balanced trade-offs, such as ensuring dataset completeness while maintaining accuracy and deciding between automation for efficiency and manual intervention for precision. We also balanced providing in-depth analysis with ensuring that results were accessible to a wider audience. Ultimately, this project not only provided valuable insights into PICR's research but also created a sustainable framework for ongoing data stewardship. The interactive dashboard and methodologies developed can be applied to other institutions, broadening the impact of our work.

If your project has published any data, please provide a URL link. N/A.

See Figure 4.2 for a screenshot of the poster titled Analyzing Research Trends.

POSTER 2'S STRENGTHS IN DATA STEWARDSHIP

ABSTRACT

- ✓ Presents examples of insights and strategies learned experientially. There is evidence of new learning about data and the knowledge generation process.

PROMPTS

- ✓ Recognizes existence of a range of trade-offs and challenges as part of data life cycle (data collection and management, consent, IP, ownership, storage).
- ✓ Provides details of a specific trade-off decision.
- ✓ Links ethics of research to data sources (privacy, participant consent, etc.).
- ✓ Shares data stewardship practices used that will help future researchers to efficiently access and use data.

POSTER

- ✓ Explains the nature of the data that led to the visualizations presented.
- ✓ Articulates the broader, societal benefits of the research (data and knowledge generated).

The strength of this case is experiential reasoning, trade-offs reasoning, and futures reasoning, with emergent elements in ethical reasoning and opportunities to grow in first-principles reasoning.

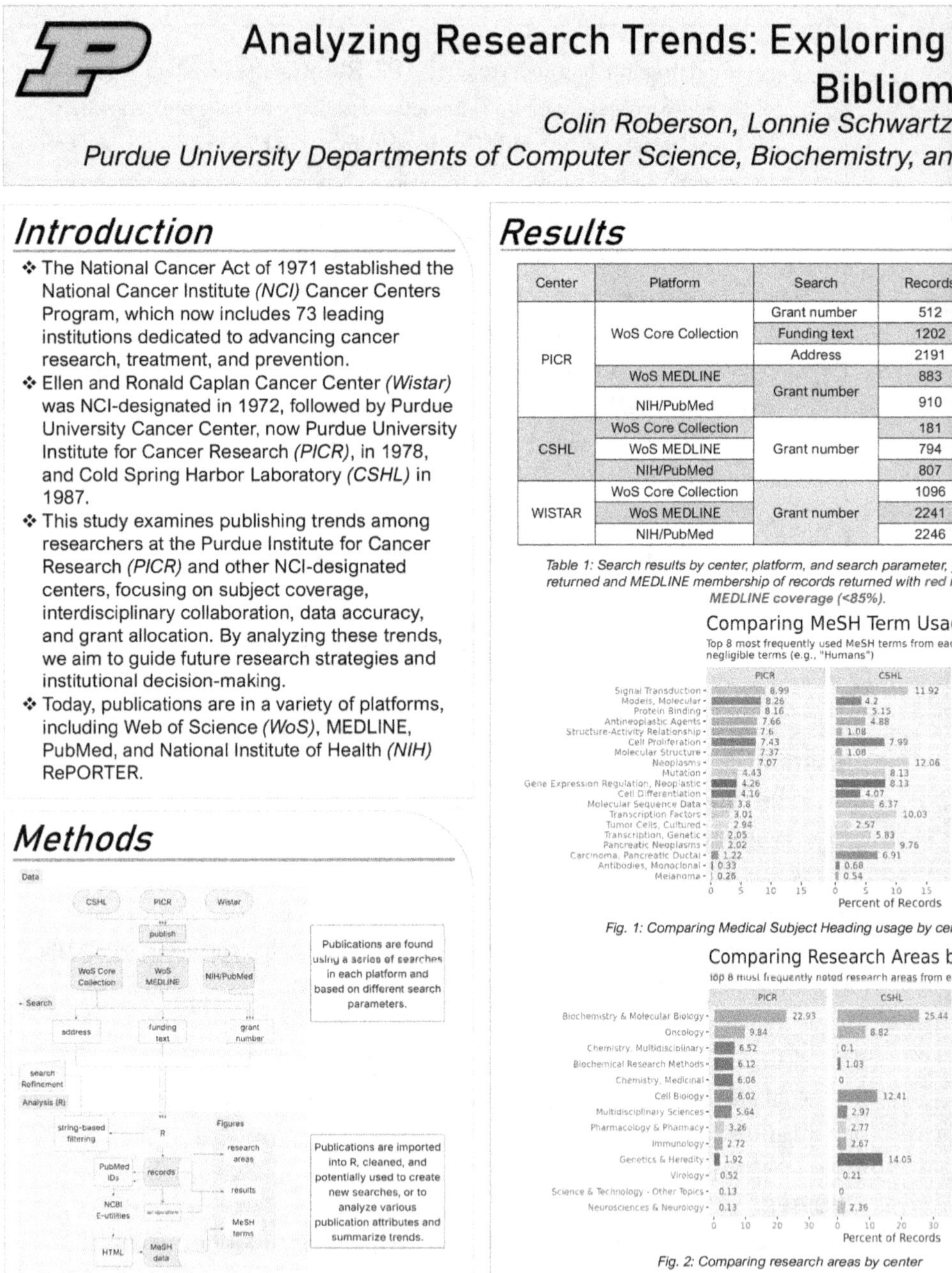

FIGURE 4.2. Poster Example 2. To view a digital version of the above poster, visit https://docs.lib.purdue.edu/stewards/2/.

l Research Impact and Expertise with ic Data

eeth Narra, and Pete E. Pascuzzi
ychological Sciences, and Libraries and School of Information Studies

ledline overage
34.38%
49.42%
21.86%
00.00%
96.62%
96.69%
00.00%
99.75%
98.54%
00.00%
99.20%

ecords lower

PICR publications from Web of Science Core Collection have noticeably less MEDLINE coverage compared to other platforms and centers.

Center
cluding

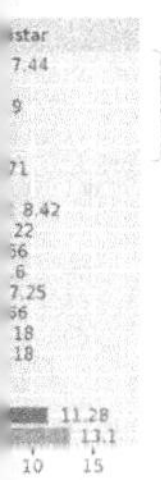

PICR's top MeSH terms are:
- ❖ Signal Transduction
- ❖ Molecular Models
- ❖ Protein Binding
- ❖ Antineoplastic Agents

er

PICR has noticeably more coverage in the following research areas:
- ❖ Chemistry, Multidisciplinary
- ❖ Biochemical Research Methods
- ❖ Chemistry, Medicinal

Conclusions

- ❖ The productivity and success of a biomedical research project can be measured by various criteria including publications, clinical trials and patents. The NIH RePORTER gathers publication data for research projects from the PubMed publication database linking to the project grant number. There are multiple issues that can cause the publication record to be incomplete including lack of journal indexing by PubMed, omission of funding details in the publication record, and human error. Our method of searching the bibliometric databases will help to address these issues.
- ❖ Preliminary analyses revealed important differences among the NCI cancer centers with respect to areas of research using both the Medical Subject Headings (MeSH) from PubMed and Web of Science Research Areas. The Purdue University Institute for Cancer Research has strengths in Antineoplastic Reagents, Molecular Structure and Molecular Models (MeSH) and the Research Areas of Medicinal Chemistry, Multidisciplinary Chemistry, and Pharmacology & Pharmacy. This reflects the strengths of Purdue University departments and faculty in these areas, suggesting that PICR makes a unique contribution to NCI mission.

References

Hall, K. L., Stokols, D., Stipelman, B. A., Vogel, A. L., Feng, A., Masimore, B., Morgan, G., Moser, R. P., Marcus, S. E., & Berrigan, D. (2012). Assessing the value of team science: A study comparing center- and investigator-initiated grants. American Journal of Preventive Medicine, 42(2), 157–163. https://doi.org/10.1016/j.amepre.2011.10.011

Stout, N. L., Alfano, C. M., Belter, C. W., Nitkin, R., Cernich, A., Lohmann Siegel, K., & Chan, L. (2018). A Bibliometric Analysis of the Landscape of Cancer Rehabilitation Research (1992-2016). Journal of the National Cancer Institute, 110(8), 815–824. https://doi.org/10.1093/jnci/djy108

POSTER EXAMPLE 3

The Statistics Behind Fantasy Basketball: A Deep Dive Into Category Leagues Team

By Karthik Selvaraj, mentored by Lisa Lynn Hayes

Abstract

Daily fantasy sports (DFS) is a great way to engage with friends and connect over a common interest. DFS is an exciting hobby where participants draft players from a professional sports league to compete with other custom-drafted teams each week. In points-based leagues, players' box score stats are plugged into formulas to generate their earned "fantasy points." By the end of the week, the team with the most fantasy points wins. However, fantasy basketball allows users to score differently in "category leagues." In category leagues, fantasy teams aggregate individual stats to try and win the most categories. As a result, ranking and drafting players become much more complex. How do you compare players that specialize in different stats? This paper aims to review and analyze the approaches to drafting and ranking players in category leagues. We compare and contrast the methods used in point and category leagues, demonstrating the unexpected complexity behind category-based fantasy basketball. While most DFS participants like to use Z-scores for their analysis, we have found other statistical methods lead to more accurate, meaningful results such as the "G-score" and "H-score." We explore how these statistical tools along with machine learning methods can yield a more insightful analysis compared to traditional rankings from DFS providers. The goal of this paper is to provide participants with more robust tools for evaluating players in category-based fantasy basketball leagues. Future work may include how these methods can be applied to different types of leagues such as dynasty or rotisserie formats.

Prompts

How did you approach decisions related to data throughout your applied research project? Basketball data is widely available online and there do exist some APIs that allow users to access this data. However, these APIs do not always provide advanced statistics and more complicated data points such as shot charts. I decided to create and compile my own SQL database with all of this data so I can structure the dataset to my needs, compile information from multiple sources, and allow for opportunities to expand the dataset at any time based on future work. For example, I am interested in also exploring how players' performance in the NCAA/EuroLeague/NBA G League/NBA Drew League affects their performance in the NBA. With my own SQL database, I can now add a "league table" to organize all of this data. Additionally, I can now associate a player's table of stats and game information with the table of the league they were in before entering the NBA. There are many possibilities now with this approach, such as creating a referee table and connecting each referee with the games they oversaw to draw insight from how they may affect game outcomes.

What considerations and trade-offs were involved in your decisions? The largest trade-off here is how much time it would take to compile all of this data into the database. I could very easily use APIs to access more basic information and conduct my analysis from this. However, this will allow me to collect and store advanced statistics and data from many seasons. This will greatly benefit my plans for future work on analyzing NBA data, so this was worthwhile. Thankfully, NBA data is publicly available, so there were not too many ethical concerns to consider. The biggest ethical concern I faced was accessing websites through web scraping and purposefully waiting to obtain information to allow others time to access information as well.

If your project has published any data, please provide a URL link. I have not published the data for this project yet, as I am still working on web scraping all of the information I want and need. The database is currently on my local network, but I will publish the dataset once I have finished collecting the data.

See Figure 4.3 for a screenshot of the poster titled The Statistics Behind Fantasy Basketball.

POSTER 3'S STRENGTHS IN DATA STEWARDSHIP

ABSTRACT

- ✓ Presents examples of insights and strategies learned experientially. There is evidence of new learning.
- ✓ Shares broader knowledge and data stewardship considerations such as data quality.

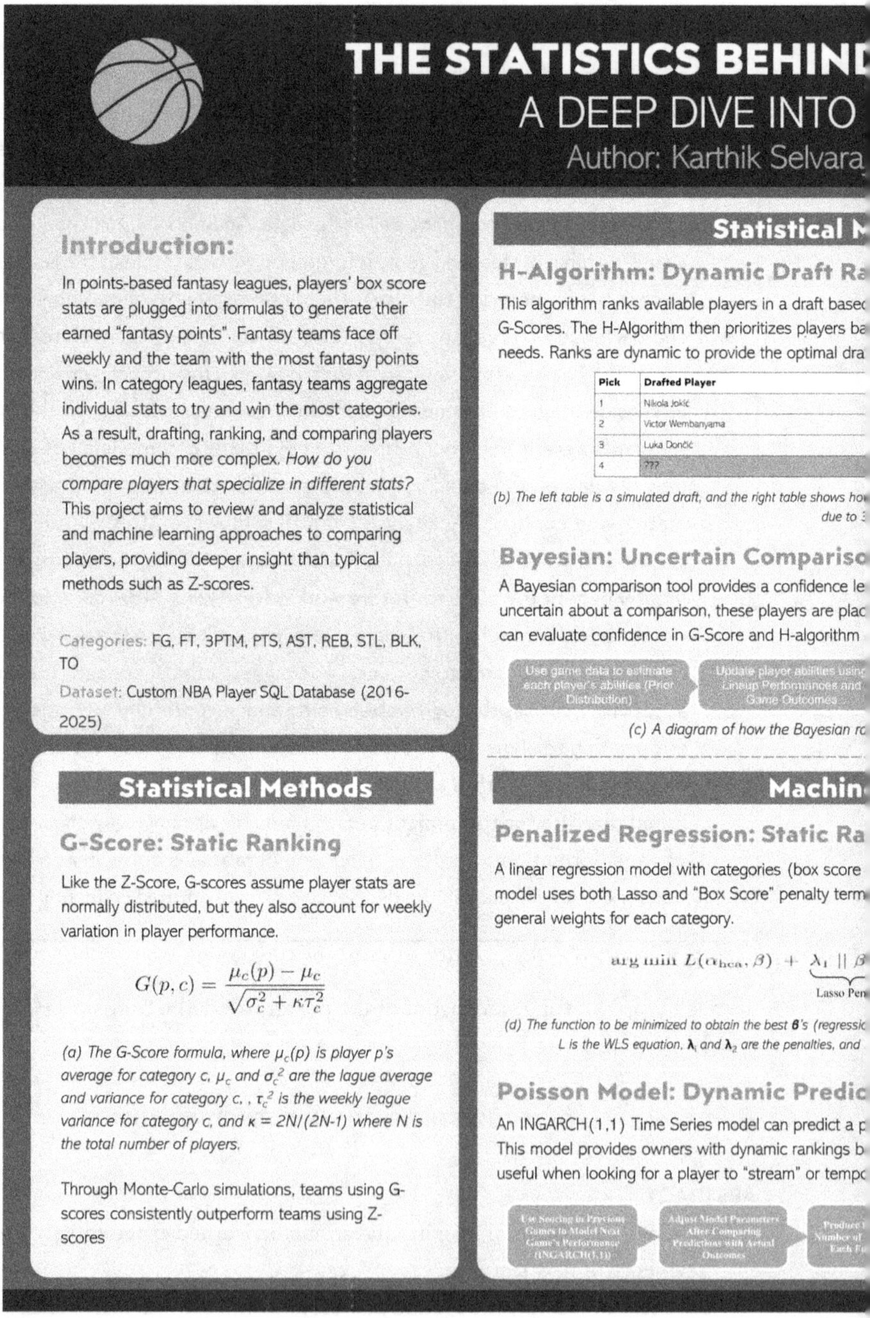

FIGURE 4.3. Poster Example 3. To view a digital version of the above poster, visit https://docs.lib.purdue.edu/stewards/3/.

FANTASY BASKETBALL:

ATEGORY LEAGUES

Mentor: Linda E Haynes

hods (cont.)

s

revious picks. Players are first ranked with modified
on the positions and categories the fantasy team
ategy.

ank	Available Player
	Shai Gilgeous-Alexander
2	Giannis Antetokounmpo
	~~Shai Gilgeous-Alexander~~
	...

l-algorithm adjusts ranks based on previous picks. SGA is boosted steals.

th its comparison prediction. When the model is
a tier together among the other rankings. This model
ngs, providing deeper insight.

Create ranges of ranks/values for players (Posterior Distribution) → Compare players in similar ranges to establish tiers

s and comparison tool are generated

earning

ng

) as predictors can generate rankings. This regression
e final output is a ranking of players for the day with

$$+ \lambda_2 \underbrace{|| \beta - z_0 \mathbf{1}_p - Rz ||_2^2}_{\text{Box Score Penalty}}$$

ficients), where α_{hca} is a learned home court advantage adjuster, Rz weights box score stats for all games for each team.

n

's expected points (and other categories) in a game.
on players' projected output. This can be especially
add to benefit from a big game.

(e) A diagram of how the Self-Exciting Poisson model generates point predictions. This can be modified to apply to other categories, as well

Machine Learning (cont.)

Principle Component Analysis (PCA)

PCA generates player comparison features, which can help with replacing injured players.

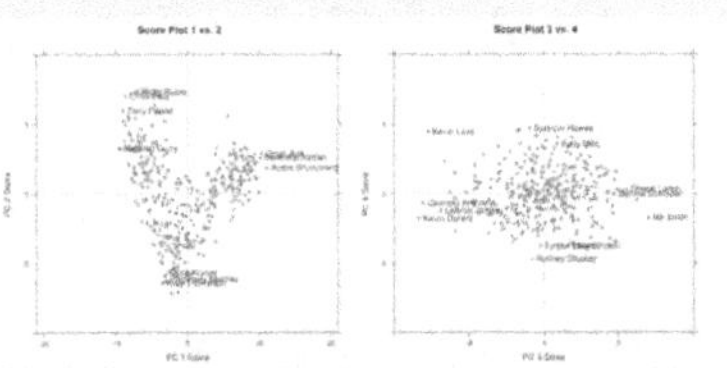

(f) Scatter plot of players using PCA features. Players close together are similar.

Conclusion:

Both statistical and machine learning methods provide fantasy owners with robust tools to analyze players and make informed, data-driven decisions. These models aid in valuing players in category-based fantasy basketball leagues the most when used together to draw insights.

The G-Score and Regression methods outperform the Z-Score for static ranking. The H-Algorithm is specialized for drafting scenarios, and the Self-Exciting Poisson model also provides dynamic rankings based on mid-season projections. Bayesian Inference and PCA are great tools for comparing similar players (find injury replacements).

Future Work:

All inclusive analysis tool, expand Poisson model to other categories, examine how NCAAB performance affects rookies for dynasty formats.

References and Citations:

→

PROMPTS

- ✓ Recognizes trade-offs in data acquisition and management.
- ✓ Demonstrates experiential reasoning in data stewardship decisions, with emergent attention to ethical considerations around responsible data access and future reuse.

POSTER

- ✓ Combining statistical and machine learning methods provides stronger, data-driven insights for valuing players in category-based fantasy basketball leagues.
- ✓ Articulates trade-offs, experiential reasoning, and broader benefits and future of the work.

The strength of this case is experiential reasoning, trade-offs reasoning, and first-principles reasoning, with emergent elements in ethical reasoning and opportunities to grow in futures reasoning.

4.4 CONCLUSION

The examination of undergraduate research posters demonstrates student learning and highlights the students' diverse data-related assets. The AI aspects of these projects, including machine learning and computer visions, demonstrate how undergraduate researchers can engage with sophisticated research. Applying sound data stewardship practices strengthens applications of data-related first principles and reasoning from experiential learning. The data stewardship framework guides these efforts with a toolkit that makes resources accessible and adaptable to different stages of research.

5

DATA MANAGEMENT TOOLKIT AND EXAMPLES

5.1. SUMMARY

The goal of this chapter is to provide data management guides, or a toolkit, that can be used in undergraduate research and other research to manage data based on the principles of data management, ethics, and stewardship. This chapter presents a range of tools (see Table 5.1) that guide data practices such as data planning, file naming conventions and organization, metadata creation, data quality assessment, storage and backup strategies, data origins, access policies, preservation protocols, repository selection, and data reuse. For most of the guides presented, one to two examples are provided to illustrate how the tools can be properly used.

5.2. DATA MANAGEMENT PLANNING DISCUSSION GUIDE

This discussion guide includes five essential components of the data management plan (DMP), built upon the data management guidelines of the National Science Foundation. The goal of this activity is to guide student teams in planning data management protocols for their research projects. Data management is an iterative process; therefore, rather than answering all the questions in the checklist, the purpose of the student team discussion is to reach a consensus and identify the remaining questions to close data management gaps throughout future iterations of the data management process over the duration of the project.

Note: The following checklist will result in a guide to the decisions made by the team to meet the goal of effective data management. This tool can be used alone or in conjunction with the other tools in this book.

TABLE 5.1. *List of Tools and Their Descriptions*

SECTION	NAME	BRIEF DESCRIPTION
5.2	Data Management Planning Discussion Guide	Provides a summary or checklist to help research team reach consensus and implement good data stewardship practices.
5.3	Data Quality Checklist	Provides a checklist to assess data quality across 6 key dimensions: data credibility, timeliness, completeness, accuracy, consistency, and deduplication.
5.4	File-Level Metadata: Data Dictionary Creation Guide	Provides a data dictionary template to help you document sufficient information about your dataset at the file level.
5.5	Project-Level Metadata: README Creation Guide	Provides a list of core components of a README template to help you document essential information about your dataset at the project level.
5.6	File Naming Convention Development Guide	Provides best practices for creating clear, consistent, and descriptive file names.
5.7	File and Folder Organization Guide	Provides tips and examples for organizing files and undergraduate research team project folders.
5.8	Data Storage and Backup	Provides guidance on selecting data storage solutions and the 3-2-1 backup data strategy.
5.9	Data Preservation Protocol	Provides guidance for preserving data in an organized and consistent manner.
5.10	Data Repository Selection Guide	Provides recommendations for choosing appropriate data repositories to publish data.

1. Types of data (experimental, computational, text-based, samples, physical collections, models, software, curriculum materials, and other materials) to be produced during the project
   - ☐ What data has your team created/used? What data will your team create/use?
   - ☐ What will be the size of the data and speed of data growth?
   - ☐ What file formats (e.g., data tables, image files, software code, XML files, websites, text documents, or physical materials) will you be using?
   - ☐ Are these formats sustainable in the long term?

2. Metadata and data description
   - ☐ How will your team document the data (i.e., provide metadata)? Metadata is data that describes collected data and provides context for understanding and reuse.
   - ☐ How will your team name and organize your files?
   - ☐ How will you ensure consistency across your team?
   - ☐ Do you have standards for data description?
   - ☐ Are you using any data dictionaries created by your team?
3. Access and sharing of data when accounting for the appropriate protection of privacy, confidentiality, security, intellectual property, export control or other requirements and ethical considerations
   - ☐ Could the data lead to a patent or other intellectual property claim?
   - ☐ What security, confidentiality, and regulations apply to the data (HIPAA, FERPA, etc.)?
   - ☐ Who will have access to your data during and after the research process? Do you have data use agreements?
   - ☐ What data will your team publish?
   - ☐ What might be the consequences of inappropriate use or access of this data?
4. Reuse, redistribution, and the production of derivatives
   - ☐ Who can reuse your data and for what purpose?
   - ☐ What are the reuse permissions for the shared data?
   - ☐ How should the data be cited when used by others?
   - ☐ Will research code be shared? If so, where will it be stored and indexed, and what reuse permissions will it have?
5. Archiving
   - ☐ How will your team back up and preserve the data?
   - ☐ What file formats will be used for archiving?
   - ☐ Where will the data be archived?
   - ☐ Who will be responsible for long-term data stewardship?
6. Notes:

5.2.1. DATA MANAGEMENT PLANNING DISCUSSION GUIDE—EXAMPLE 1

In this example, one undergraduate research team spent approximately 15 minutes at the beginning of the semester taking notes to document their data workflows (see Figure 5.1). This helped them identify gaps within their data stewardship and

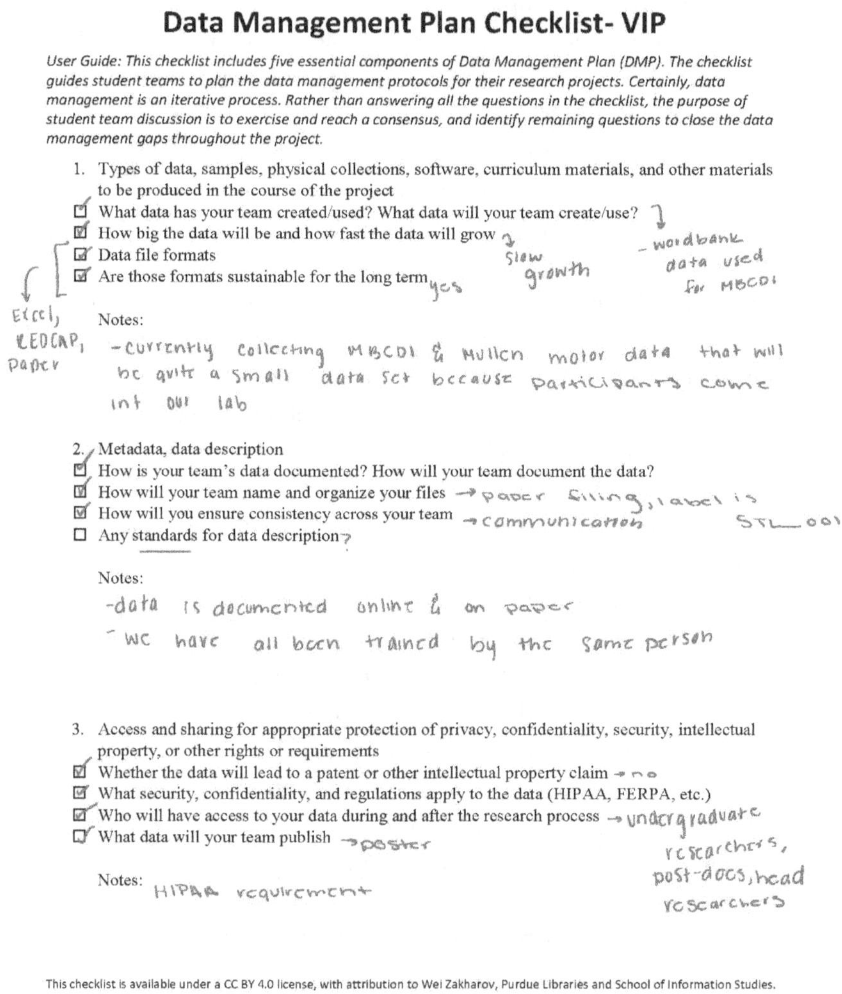

Data Management Plan Checklist- VIP

User Guide: This checklist includes five essential components of Data Management Plan (DMP). The checklist guides student teams to plan the data management protocols for their research projects. Certainly, data management is an iterative process. Rather than answering all the questions in the checklist, the purpose of student team discussion is to exercise and reach a consensus, and identify remaining questions to close the data management gaps throughout the project.

1. Types of data, samples, physical collections, software, curriculum materials, and other materials to be produced in the course of the project
   - ☑ What data has your team created/used? What data will your team create/use? — wordbank data used for MBCDI
   - ☑ How big the data will be and how fast the data will grow — slow growth
   - ☑ Data file formats — Excel, REDCAP, paper
   - ☑ Are those formats sustainable for the long term — yes

   Notes:
   - currently collecting MBCDI & Mullen motor data that will be quite a small data set because participants come into our lab

2. Metadata, data description
   - ☑ How is your team's data documented? How will your team document the data?
   - ☑ How will your team name and organize your files → paper filing, label is STL_001
   - ☑ How will you ensure consistency across your team → communication
   - ☐ Any standards for data description?

   Notes:
   - data is documented online & on paper
   - we have all been trained by the same person

3. Access and sharing for appropriate protection of privacy, confidentiality, security, intellectual property, or other rights or requirements
   - ☑ Whether the data will lead to a patent or other intellectual property claim → no
   - ☑ What security, confidentiality, and regulations apply to the data (HIPAA, FERPA, etc.)
   - ☑ Who will have access to your data during and after the research process → undergraduate researchers, post-docs, head researchers
   - ☑ What data will your team publish → poster

   Notes: HIPAA requirement

FIGURE 5.1. Data management plan checklist completed by undergraduate research team.

4. Re-Use, re-distribution, and the production of derivatives
- ☑ Who can re-use your data and for what purpose
- ☑ What are the re-use permissions for the shared data → reuseable
- ☑ How should the data be cited → ?
- ☑ Whether and where research code will be shared and under what reuse permissions → ?

Notes: data we are using is part of a larger study

5. Archiving
- ☑ How will your team back up and preserve the data → backed up in redcap
- ☑ What file formats will be used for archiving → paper
- ☑ Where the data will be archived → in lab, online
- ☑ Who will be responsible for long-term data stewardship ↓ head researcher in lab

Notes:

* I don't know much about archive process

FIGURE 5.1. *(Continued)*

assisted with the onboarding process for new students as they discussed their team's current data organization.

5.2.2. DATA MANAGEMENT PLANNING DISCUSSION GUIDE—EXAMPLE 2

In this example, the undergraduate research team chose a lengthier method. Rather than filling out the discussion guide, they developed their own data management plan document, guided by the contents of the discussion guide.

> Our team created a big range of data, focusing on physical structures and their properties like temperature and pressure through different test runs, which include both experimental results and computational simulations. Moving forward, we plan to expand our dataset to include more complex computational models and graphics and the effects of those changes on the properties of the original datasets we possess right now. The size of our data is expected to grow significantly (with the total capacity potentially exceeding 1.5 terabytes) as we delve into more detailed analyses due to the high-resolution nature of the raw and processed data involved. This growth will be gradual (approximately half a terabyte a year is projected) but steady as our experiments and simulations become more complex.
>
> Currently, our data is stored in formats such as CSV for raw numerical data, PNG for visualizations, and proprietary formats like PDB and CAD for 3D modelled structures. While these formats have served us well, we're exploring more sustainable options for long-term storage and accessibility, primarily due to incompatibility of several CAD formats and lack of interchangeability of tools between different CAD software. The solutions include formats like IGES and STL for large numerical datasets and standardized formats for physical data that ensure compatibility and ease of use across different computational platforms and databases. Our goal is to ensure that our data remains accessible, usable, and secure for years to come, facilitating not only our current research but also enabling future studies.
>
> In our research, we practice data documentation and organization to easily reach and communicate our findings. Our current approach involves metadata

files for each dataset we obtain or process, capturing essential information like experimental conditions, pressure variations, and analytical methods used. We plan to refine our documentation by adopting structured metadata standards, including the Dublin Core schema, to provide a foundation for metadata and for accessible technical details. Our file naming and organization strategy relies on a clear convention and structured directory system, categorized by project phase, and experiment type. Consistency is ensured through weekly meetings and standardized templates, supplemented by version control.

This commitment to meticulous data management practices makes sure that our project is transparent and reproducible, aligning with both our immediate project needs and the broader scientific community's goals.

Given the context of our research project and its academic nature, we do not anticipate filing for patents or other intellectual property claims at this stage due to the preliminary and yet unproven nature of our results. Our project operates without the constraints of security or confidentiality regulations such as HIPAA or FERPA, as it does not involve sensitive personal information or educational records.

Throughout the duration of our research and thereafter, we are committed to an open access policy.

This means that our data, which will include models, drawings, experimental setups, simulation results, and sample properties, will be freely accessible to the public. This approach aligns with our goal of contributing to the collective knowledge within our field and fostering collaboration and innovation.

Following the conclusion of our research, the results and associated data will be published and made available to the wider academic community and the public. Our intent is to publish all aspects of our work.

The research data and code are available for reuse by anyone for any purpose, aiming to promote further scientific research. We plan to license our data under a Creative Commons CC BY license, and our research code will be shared on platforms like GitHub under open-source licenses such as the MIT License or GPL. These choices ensure that our work can be freely used, adapted, and shared, provided proper attribution is given and the open-sharing principles are maintained.

To ensure the integrity and longevity of our data, we'll implement a dual backup strategy using local and cloud storage, with routine scheduled backups.

Our choice of file formats for archiving (CSV, TXT, and PDF for general data, and discipline-specific standards for specialized data) reflects our interests in accessibility and compatibility. Our data will be archived in institutional databases, offering secure, long-term storage while ensuring our research is accessible. For broader dissemination, platforms like Zenodo or Dryad may be considered. A dedicated team member will be responsible for the stewardship of our data. This includes overseeing data management practices, maintaining the integrity of our backups, and ensuring our data remains accessible and up-to-date with evolving technology standards.

5.3. DATA QUALITY CHECKLIST

A data quality checklist was developed to facilitate undergraduates' understanding of the critical aspects of data quality assessment. This data quality checklist encompasses six fundamental dimensions: data credibility, timeliness, information completeness, data accuracy, data consistency, and data deduplication. Each dimension plays a crucial role in ensuring the reliability and usability of a dataset.

DATA CREDIBILITY

- ☐ Is information available regarding who created the data?
- ☐ Is information available regarding who published the data?
- ☐ Is information available regarding who contributed to the data?
- ☐ Is contact information available?

TIMELINESS

- ☐ Are the data previously collected still current?
- ☐ Is information available regarding the last time data was captured or updated?
- ☐ Is version control implemented to track revisions in the dataset?

INFORMATION COMPLETENESS

- ☐ Have you checked for missing elements in the data records?
- ☐ Have you checked for missing data records?

- ☐ Has all information been captured for their intended use?
- ☐ Are the data findable and reusable by someone inexperienced with the dataset?

DATA ACCURACY

- ☐ Have you fixed all data typos?
- ☐ Are all the data formats correct?
- ☐ Have you accounted for data outliers that may not be recorded accurately?
- ☐ Do the data represent the information we intend to capture?

DATA CONSISTENCY

- ☐ Are the data formats consistent?
  - ☐ Referring to how a value is displayed, such as the number of decimal places.
- ☐ Are the data types consistent?
  - ☐ Referring to the underlying type of the data, such as integers, decimals, text, and lists.
- ☐ Are the data unit measurements consistent?
  - ☐ Referring to whether measurement units (e.g., length in inches, currency in USD, weight in pounds) are used consistently across the dataset.
- ☐ Have data points been synched within and across platforms?

DATA DEDUPLICATION

- ☐ Have repeated data records been removed and/or accounted for?

5.3.1. DATA QUALITY CHECKLIST—EXAMPLE 1

In this example, the undergraduate team addressed project questions such as "What are the impacts of Covid vaccines on the new case numbers released for Tennessee?" and "Are people of a particular age more vulnerable to Covid?" After selecting two data sources, the students assessed the data they were planning to use for their project to ensure that it was of the quality specified by the Data Quality Checklist.

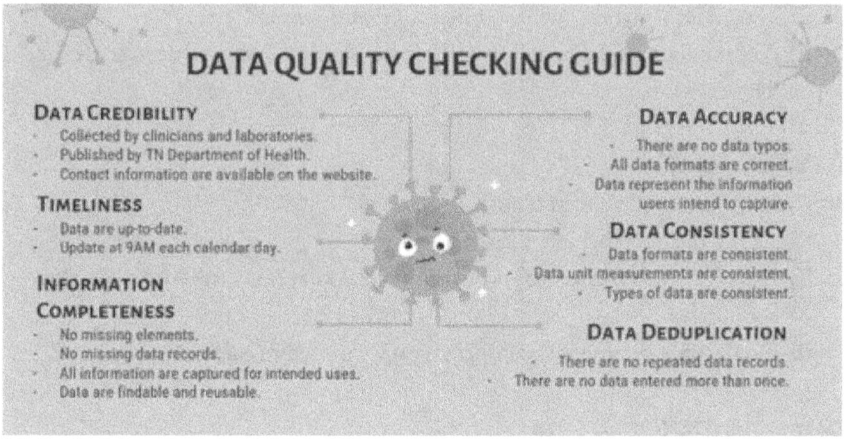

FIGURE 5.2. Data quality checking guide created by project team.

PROJECT QUESTIONS

- What are the impacts of Covid vaccines on the new case numbers released for Tennessee?
- Are people of a particular age more vulnerable to Covid?

EXTERNAL DATA SOURCES

- TN Department of Health. (2021). *Daily case information* [Data file]. Retrieved from https://www.tn.gov/health.html
- TN Department of Health. (2021). *COVID vaccine state summary* [Data file]. Available at https://www.tn.gov/content/dam/tn/health/documents/cedep/novel-coronavirus/datasets/COVID_VACCINE_STATE_SUMMARY.XLSX

See Figure 5.2 for the data quality checking guide answers.

5.3.2. DATA QUALITY CHECKLIST—EXAMPLE 2

In this example, the undergraduate team addressed project questions such as "How do different driving scenarios impact driver's choice of accepting paper coupons by stores?" After selecting a data source, the students assessed the dataset

DATA QUALITY CHECKING GUIDE ANSWERS

DATA CREDIBILITY	TIMELINESS	INFORMATION COMPLETENESS	DATA ACCURACY	DATA CONSISTENCY	DATA DEDUPLICATION
Who created the data? **Tong Wang** **Cynthia Rudin**	Are data outdated? **Created on 9/15/2020**	Are there missing elements in data records? **Yes**	Are there data typos? **No**	Are data formats consistent? **Yes**	Are there repeated data records? **No**
Who published the data? ***The Journal of Machine Learning Research***	When are the data captured and updated? **A Survey on Amazon Mechanical Turk**	Are there missing data records? **Yes**	Are data formats correct? **Yes**	Are data unit measurements consistent? **Yes**	Are data entered more than once? **No**
Who contributed to the data? **Finale Doshi-Velez** **Yimin Liu** **Erica Klampfl**	Is version control implemented to track revisions of a dataset? **No, 1-time survey**	Is all information captured for its intended use? **Yes**	Are there data outliers which may not be recorded accurately? **No**	Are types of data consistent? **Yes**	
Is contact information available? **Yes**		Are data described to be findable and reusable? **Yes**	Do data represent the information we intend to capture? **Yes**	Are data synched within and across platforms? **Yes**	

they were planning to use for their project to ensure that it met the criteria specified by the Data Quality Checklist.

PROJECT QUESTIONS

- How do different driving scenarios impact driver's choice of accepting paper coupons by stores? Investigated factors include age group, weather, existence of a passenger, and income.

EXTERNAL DATA SOURCES

- UC Irvine Machine Learning Repository. Data was collected via a survey on Amazon Mechanical Turk.

5.4. FILE-LEVEL METADATA: DATA DICTIONARY CREATION GUIDE

First, what is metadata? Metadata is data that describes collected data and provides context for understanding and reuse. For example, when you take a photo, the photo itself is the collected data, while the properties (such as time taken, byte size, and file format type) are the metadata. In research, metadata is created and maintained to facilitate the sharing and reuse of data because it provides important information regarding the circumstances behind the data and helps other researchers determine whether the data are fit to be used for their projects.

ELEMENTS TO INCLUDE

While more information can be added, at a minimum, the metadata elements identified in Figure 5.3 should be included.

STANDARDIZATION

To maintain clarity within a project, a standardized metadata schema should be established. For many fields of study there exist predefined metadata standards, such as the general purpose Dublin Core schema or the biology specialized Darwin Core schema. These predefined standards can make it easier to apply metadata to project datasets. Regardless of which schema is chosen, you should include a file to describe the standards used in your work.

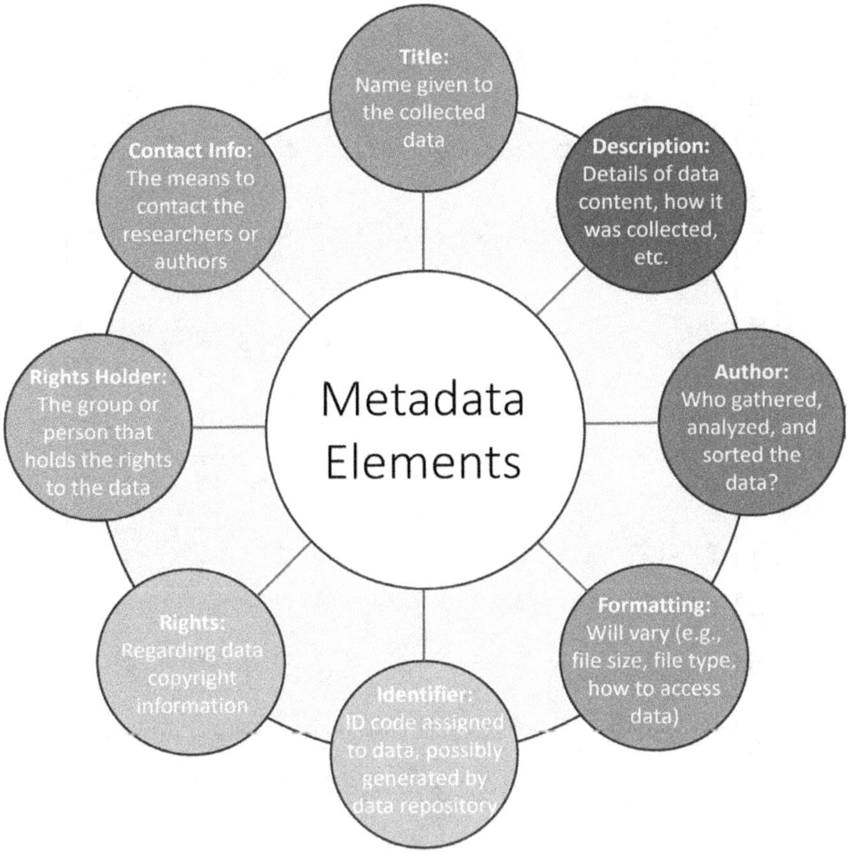

FIGURE 5.3. Collection of important metadata elements used in data documentation.

5.4.1. FILE-LEVEL METADATA—DATA DICTIONARY EXAMPLE 1

In this example, a list of dairy food products is organized by nutritional value and serving size measurements. The sample dataset was obtained from Kaggle, an online platform where massive public datasets are available. This specific dataset was listed under food nutrition analysis for common food nutritional facts. This data table includes rows representing an individual food item and columns representing the measurement, grams, calories, protein, fat, saturated fat, fiber, carbs, and category.

SAMPLE DATA TABLE

FOOD	MEASURE	GRAMS	CALORIES	PROTEIN	FAT	SAT. FAT	FIBER	CARBS	CATEGORY
Cows' milk	1 qt.	976	660	32	40	36	0	48	Dairy products
Buttermilk	1 cup	246	127	9	5	4	0	13	Dairy products
Evaporated milk, undiluted	1 cup	252	345	16	20	18	0	24	Dairy products
Fortified milk	6 cups	1,419	1,373	89	42	23	1.4	119	Dairy products
Powdered milk	1 cup	103	515	27	28	24	0	39	Dairy products
Skim milk, instant	1⅓ cups	85	290	30	t	t	0	42	Dairy products
Skim milk, non-instant	⅔ cup	85	290	30	t	t	1	42	Dairy products
Goats' milk	1 cup	244	165	8	10	8	0	11	Dairy products
Cocoa	1 cup	252	235	8	11	10	0	26	Dairy products
Skim milk	1 cup	250	128	18	4	3	1	13	Dairy products
Custard	1 cup	248	285	13	14	11	0	28	Dairy products
Ice cream	1 cup	188	300	6	18	16	0	29	Dairy products
Ice milk	1 cup	190	275	9	10	9	0	32	Dairy products
Cream or half-and-half	½ cup	120	170	4	15	13	0	5	Dairy products
Whipping cream	½ cup	119	430	2	44	27	1	3	Dairy products
Cheese	1 cup	225	240	30	11	10	0	6	Dairy products

The data dictionary for the sample table lists the following variables: food, measure, grams, calories, protein, fat, saturated fat, fiber, carbs, and category. For each variable, the data dictionary provides a description, data type, unit of measurement, decimal precision (if applicable), and any additional notes needed to clarify how the variable should be interpreted or used.

5.4.2. FILE-LEVEL METADATA—DATA DICTIONARY EXAMPLE 2

This example shows the type of information that is typically recorded in an interview dataset. In particular, the example illustrates how metadata are structured for audio-based interviews that have been transcribed to text.

5.5. PROJECT-LEVEL METADATA: README CREATION GUIDE

Project-level metadata describes the who, what, when, where, and why of a dataset or project folder. A common example is the README file, which provides essential information for anyone accessing a dataset for the first time. The README file provides context, such as project descriptions, file organization, authorship, usage guidance, and other key details needed to understand the dataset as a whole. Regarding the README file type, plain text or Markdown file formats are recommended for longevity.

README TEMPLATE EXAMPLES

- Cornell University's README template guide, https://data.research.cornell.edu/data-management/sharing/writing-readmes-for-research-code-software/; https://data.research.cornell.edu/data-management/sharing/readme/
- GitHub's curated awesome-README list, https://github.com/matiassingers/awesome-readme

DATA ADOPTION CERTIFICATE TEMPLATE

One way to utilize metadata is to use it as an adoption certificate for a dataset. An adoption certificate promotes owner attribution, proper file naming conventions, documentation, and other essential metadata elements. A certificate is also

DATA DICTIONARY TABLE

VARIABLE	DESCRIPTION	DATA TYPE	UNIT	PRECISION	COMMENT
Food	Food item name	Categorical	N/A	N/A	The letter *t* was used to indicate a miniscule amount.
Measure	Amount of food analyzed per serving	Numerical, continuous	Variable	Integer	
Grams	Amount the food item weighed	Numerical, continuous	grams	Integer	
Calories	Amount of calories food item contains	Numerical, continuous	kcal	Integer	
Protein	Amount of protein food item contains	Numerical, continuous	g	Integer	
Fat	Amount of fat food item contains	Numerical, continuous	g	Integer	The letter *t* was used to indicate a miniscule amount.
Sat. Fat	Amount of saturated fat food item contains	Numerical, continuous	g	Integer	The letter *t* was used to indicate a miniscule amount.
Fiber	Amount of fiber food item contains	Numerical, continuous	g	.0	
Carbs	Amount of carbs food item contains	Numerical, continuous	g	Integer	The letter *t* was used to indicate a miniscule amount.
Category	Defines the food item being described	Categorical	N/A	N/A	Only dairy products were observed.

DATA DICTIONARY TABLE

INTERVIEW NUMBER: 00-1	
GENERAL INFORMATION	
Contributor Information	**Name of Person Interviewed:** Jane Doe
	Major: Engineering
	Gender: Female
Date	**Day:** 2024-02-14
	Time: 2:35 PM to 3:15 PM (extra time was spent with suggestions and thoughts)
	Location: WALC 3053
Creator	**Name of Interviewer:** John Nickle
Coverage	**Subject of Interview:** Data management in undergraduate research
Description	**Description of Interview Topic:** An in-person interview given by John Nickle to the person interviewed, Jane Doe, about the use of data in past undergraduate research projects
Relation	**Relation:** Coworkers
METADATA INFORMATION	
Format	**Type of Recording Data:** Transcript
	Language: English
	Publisher: Not published
	Source of Data: Interviewee (Jane Doe); interviewer (John Nickle)
Type	**Type:** Academic research interview
	Undergraduate? Yes
	In-Person? Yes
	Number of Interviewed Attendees: 1
Title	**Interview:** 00-1
RIGHTS PROVIDED TO THOSE INTERVIEWED	
Rights	We will not disclose your information to external parties without direct permission.

CORE COMPONENTS OF THE README TEMPLATE (DATASET-FOCUSED)

COMPONENT	DESCRIPTION
Dataset/project title	Clear and concise name of the dataset or research project.
Dataset description	A brief overview of what the dataset contains, its purpose, scope, and intended audience.
Table of contents	Optional, but helpful for navigation, especially for longer READMEs.
Creator(s)/contributor(s)	Names, affiliations, and roles of individuals or organizations involved in creating the dataset.
Dates	Key dates related to the dataset, including data collection period, publication date, and version dates if applicable.
Geographic and temporal coverage	Description of the spatial and temporal scope of the data, if relevant.
Data collection	How the data were generated, collected, or compiled, including instruments, protocols, or study design.
File and folder organization	Explanation of directory structure, file naming conventions, and relationships among files.
Data dictionary	Definitions of variables, units of measurement, codes, missing values, and controlled vocabularies.
Data formats	File formats used (e.g., CSV, XLSX, TXT) and any relevant standards.
Data quality and limitations	Known issues, sources of error, data cleaning steps, and limitations affecting reuse.
Access and use information	Instructions for accessing the dataset, including permissions, restrictions, or authentication if applicable.
License and reuse conditions	Terms of use, licensing information, and recommended citation for reuse.
Ethical and legal considerations	IRB status, consent considerations, anonymization/de-identification steps, and compliance notes.
Related publications and outputs	Associated articles, reports, presentations, or software linked to the dataset.
Versioning and change log	Description of dataset versions and a summary of major changes over time.
Contact information	How users can contact the dataset creators or maintainers for questions or issues.
Funding acknowledgment	List grant numbers if the project is grant-funded.

CORE COMPONENTS OF THE README TEMPLATE (SOFTWARE-FOCUSED)

COMPONENT	DESCRIPTION
Project title	Clear and concise name of your project.
Description	A brief overview of what the project does, its purpose, and who it's for.
Table of contents	Optional, but helpful for navigation, especially for longer READMEs.
Installation instructions	Step-by-step instructions to install/set up the project, including software/hardware requirements, dependencies, prerequisites, and platform-specific notes.
Usage	Examples of how to run or use the project, including code snippets, screenshots, or GIFs.
Features	Highlight key functionalities or modules.
Configuration	Details on how to customize or configure the project.
Contribution	Guidelines for how others can contribute (e.g., coding standards, pull request process).
License	Specify the license under which the project is distributed (e.g., MIT, GPL).
Authors and acknowledgements	Credit contributors, collaborators, or third-party tools.
Contact information	How to reach the maintainers or report issues.
Known issues/FAQs	List of bugs, limits, troubleshooting, or frequently asked questions.
Changelog	Summary of major updates and version history.
References/related projects	Links to documentation, research papers, or similar projects.
Funding acknowledgment	List grant numbers if the project is grant-funded.

Data Adoption Certificate

This is to certify that [data name]

____

Version # __________

Was created by

____

On this [day, month, year]

________ / ________ / ________

At location

____

With the data format

____

EXPERIMENT	SUBJECT
Experiment identifier or version number: __________	Subject of research name: __________
Experiment type: __________	Overview about the subject of research: __________
Other experiment-specific details: __________	Subject or sample source: __________

an anthropomorphism that ai ms to impose inanimate characteristics on data. Through this literacy device, we aim to make data relatable and the process of data management more desirable.

5.6. FILE NAMING CONVENTION DEVELOPMENT GUIDE

Have you ever been frustrated by having to open every file in a folder to locate the single document you need? How can this hassle be avoided? Similar to each

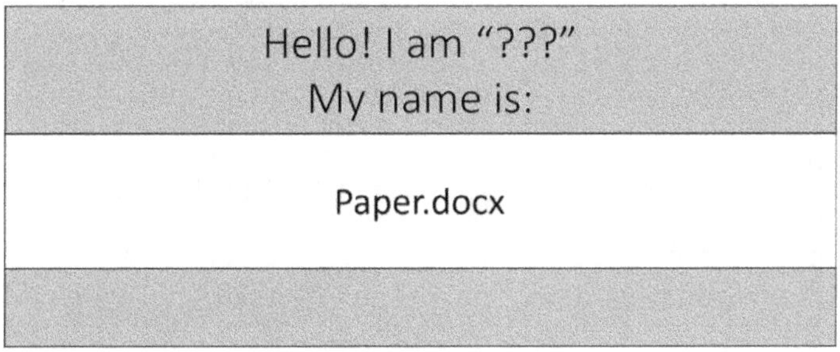

FIGURE 5.4. File name tags (uninformative).

researcher having a name, every file created requires a proper name. In fact, the file names should be more specific than our name. Imagine that your phone book includes 10 different people named Elsa. How would you know which Elsa to call? Without the last names and/or notes, this task would be difficult. Similar to the phone book example, we can consider file names as name tags. Providing only a first name would be analogous to naming a file "paper.docx."

Figure 5.4 shows an uninformative file naming example. Not very clear, is it? Very little information about a file can be obtained using this approach. By providing more specific information about a file within its name, it becomes easier to find it, either by browsing or using search functions on our devices. Additionally, a proper file naming convention allows for consistent sorting within file groups and easy clarification of a file's contents. Figure 5.5 lists meaningful file naming examples.

COMMON FILE NAMING GUIDELINES

When creating names for files, a few commonly held guidelines should be followed. By adhering to these guidelines, you will ensure that your file names are easily interpreted by both humans and machines. Figure 5.6 lists naming convention guidelines.

POTENTIAL INFORMATION TO INCLUDE IN YOUR FILE NAME

- Date (recommended format either YYYYMMDD or YYYY-MM-DD)
- Experiment type
- Subject of research name
- Creator (last, first or initials)
- Details about the subject of research
- Data format

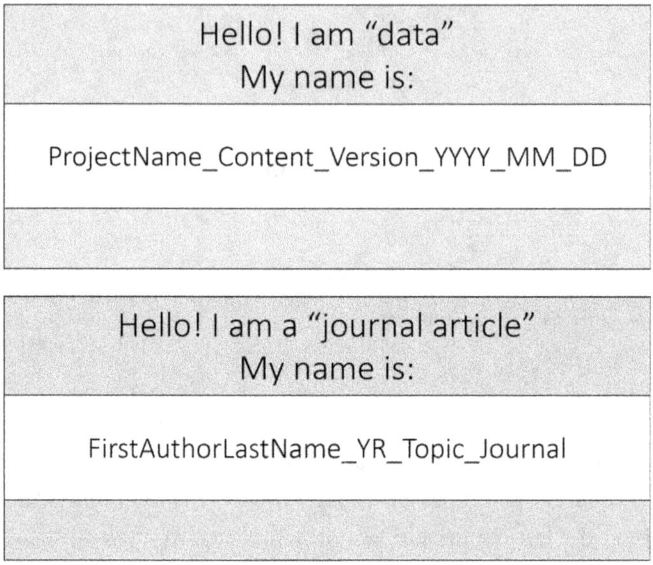

FIGURE 5.5. File name tags (informative).

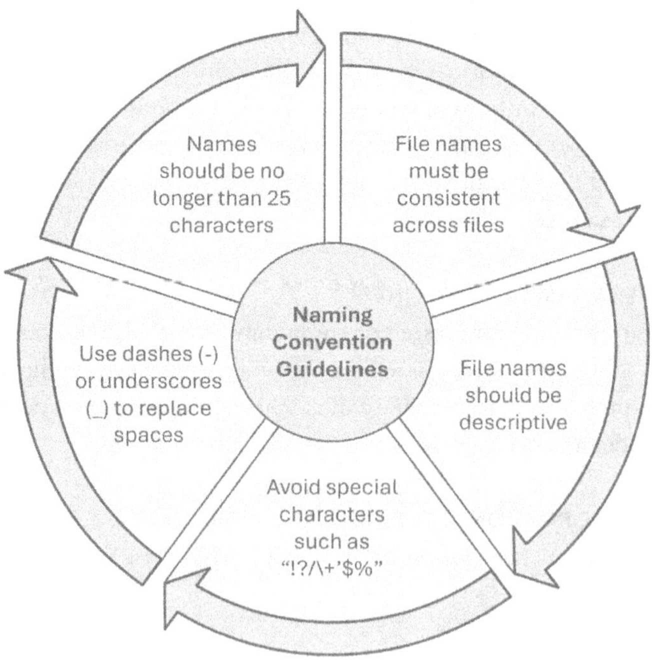

FIGURE 5.6. File naming guidelines.

- Sample source
- Experiment identifier/version number
- Other experiment-specific details

EXAMPLES OF PROPER FILE NAMING

YYYYMMDD_ExperimentType_ExperimentNumber.FileFormat
YYYY-MM-DD_SubjectOfResearch_SiteOfResearch_VersionX.FileFormat
ResearcherInitials_SampleSource.FileFormat
FirstAuthorLast_Year_topic_journal.FileFormat

EXAMPLES OF ACTUAL FILE NAMES

20240301_WindTunnel_Mach2_0trial.csv
20251218_spectroscopy_213.xlsx
Vaswani_2017_AttentionIsAllYouNeed_NeurIPS.pdf

5.7. FILE AND FOLDER ORGANIZATION GUIDE

Organizing your dataset well makes it easier and faster to find the files and data you need during a research project. Because each research project has unique workflows and priorities, organizational systems should be tailored to support the specific needs and practices of project teams. Please avoid deeply nested folder structures (more than three levels) and include visual diagrams depicting sample file and folder organization.

FOLDER CATEGORIZATION

When setting up the folder structure for your project, there are a few different categories you can use to organize data. A project could potentially store data by:

- Date and time
- Experiment type
- Name of researcher that collected the information
- Data format
- Research notebook page number
- Source of data

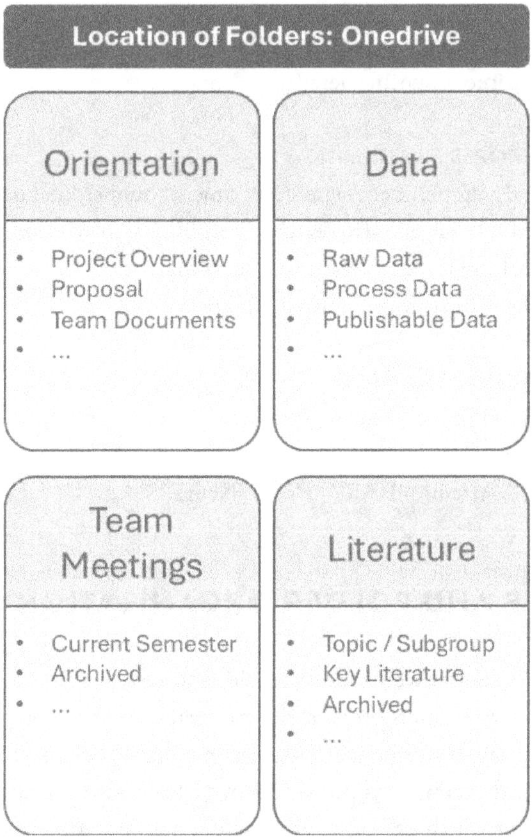

FIGURE 5.7. Folder organization example.

Each project has its own priorities regarding these categories. Many successful organizational structures use a combination of several categories.

UNDERGRADUATE RESEARCH PROJECT FILE ORGANIZATION EXAMPLE

In the example shown in Figure 5.7, each box represents a folder in the directory, with bullet points indicating the information stored within it and its organization.

Alternatively, the example project folder structure diagram that follows is commonly used to quickly illustrate a typical project folder structure.

EXAMPLE PROJECT FOLDER STRUCTURE

```
Project/
├── Orientation/
│ ├── Project overview/
│ └── Proposal/

│ └── Team organization/

├── Data/
│ ├── Raw data/
│ └── Process data/
│ └── Publishable data/
├── Team meetings/
├── Literature/
├── Results/
└── README.md
```

BUILDING AN ORGANIZATIONAL STRUCTURE

To build an organizational structure for your data files, consider the following questions:

CONTENT

- What categories of data will be generated in this project that should become folders?
- Where will metadata, documentation, and file naming conventions be stored?

DEVELOPMENT

- Will your data go through multiple stages of development, and how will your team account for that in your folder organization?
- Will you create a dedicated folder for raw data? (Raw data should always be kept separate to ensure it is never overwritten.)
- Will you need temporary storage for data that is currently being processed?

EASE OF ACCESS

- What data categories will be accessed frequently?
- How can you organize your data so these frequently accessed files can be found quickly?
- Who will be the contact person if someone needs to access the data?

MAINTAINING THE SYSTEM

- How will your team ensure that data organization is being upheld?
- When will your team update and organize files and the project folder?

5.8. DATA STORAGE AND BACKUP

Data collected over the course of a project requires storage. There are many options to choose from when storing your data. Once data are collected, it is important to decide where to store them and how to ensure that they are not lost. Choosing appropriate data storage media and backup strategies is essential for maintaining the security and accessibility of the dataset.

The following are common data storage media:

- Desktop computers and laptops
- External hard drives
- Optical discs
- Institutional and/or departmental drives
- Cloud storage such as Google Drive, Box, Dropbox, OneDrive, and so forth

Several key factors should guide the decision when considering data storage media:

1. *Storage space* is a primary consideration because it determines the volume of data that can be accommodated.
2. The *cost* of storage media is another consideration because higher-performance or higher-capacity options are generally more expensive.
3. The *intended use* of storage media should align with the intended application, such as archival purposes, frequent access, or high-performance computing (HPC) environments.

4. *Data transfer speed and latency* are critical for performance-sensitive applications, particularly for workflows involving large datasets, real-time analysis, or HPC environments where fast read/write access is essential.
5. The *lifespan* should be considered for long-term projects because some media degrade over time, making redundancy and backups essential.
6. *Compliance with institutional and funder policies* is necessary to ensure proper data management, retention, and security practices.
7. *Accessibility* is a key consideration, particularly in terms of how easily data can be retrieved and utilized by the authorized users. Compatibility with HPC systems is essential for applications that require high-speed data processing and analysis.
8. *Data ownership and licensing* must be addressed to ensure compliance with legal and ethical standards and safeguard intellectual property rights and user agreements.

DATA BACKUP 3-2-1 RULE

To properly back up open data for a project, researchers typically follow the 3-2-1 rule shown in Figure 5.8.

By retaining three copies of the data (one original and at least two copies), using different types of media for storage (such as physical, local storage, and cloud), and separating backup data from potential breaches, corruption, or hacking that can occur, you can ensure that you have recoverable copies of data if necessary.

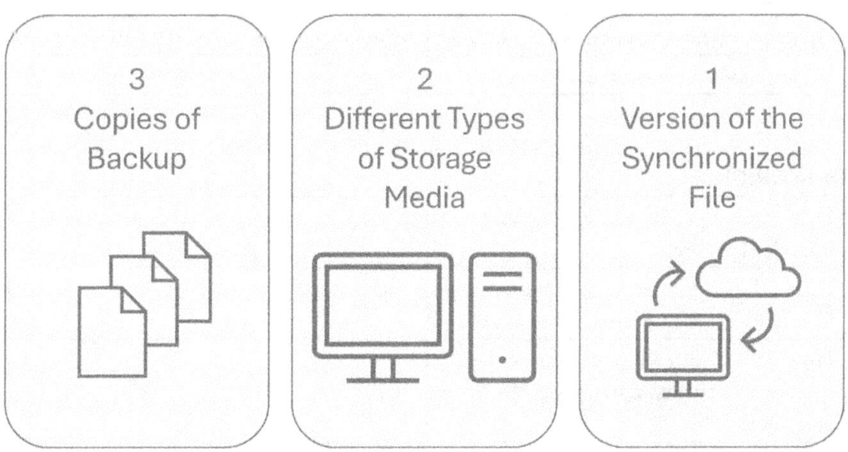

FIGURE 5.8. The 3-2-1 rule for backing up open data.

Some sensitive data cannot follow the standard 3-2-1 backup rule owing to privacy, legal, or security restrictions. Instead of multiple copies or cloud storage, these data should be stored in *encrypted, access-controlled locations*, with backups stored on *secure local or physical media*. Regular audits and monitoring ensure that data remain safe from unauthorized access, corruption, or accidental deletion. Even without full 3-2-1 redundancy, careful planning and strict security measures help maintain data integrity and availability when required.

5.9. DATA PRESERVATION PROTOCOL

Data preservation protocol provides guidance to help researchers preserve their data in an organized and consistent manner. It emphasizes that data preservation is an active, ongoing process rather than a one-time action. By keeping a record of the basic project information, data origins, access policies, and filing organization and naming conventions in a format as shown below, researchers will gain a better understanding of the organizational systems that will be used and maintained throughout the life cycle of their projects.

BASIC PROJECT INFO

Research project title	
Brief project description	
Time period for research	
Researchers	
Graduate research assistants	
Undergraduate research assistants	

DATA ORIGINS

Data sources produced by the team (e.g., interview, survey)	
External data sources (e.g., scholarly library, external interview, inherited)	
How do you differentiate external data sources and created data sources in your files?	

ACCESS POLICIES

Data access policies (Who is permitted to access your data?)	
Data preservation policies (How will you ensure that the data are properly preserved and accessible to those who require access?)	
Data transfer process (How will new members of the team gain access to old team files?)	

FILE ORGANIZATION AND NAMING CONVENTIONS (*SEE MORE DETAILS IN SECTIONS 1.5 AND 1.6*)

File organization map and location (Where can you find a map for file organization; e.g., Teams, Google Drive, Microsoft Box?)	
Naming conventions (Where can you access the naming conventions for your files?)	

5.9.1. DATA PRESERVATION PROTOCOL EXAMPLE

In this example, the undergraduate team chose to fill out the preservation protocol in paragraph form. They addressed each question properly, with reference to their own project and team members.

> We will implement cloud storage backups (Google Drive) for remote access. Regular backups will be scheduled periodically, with critical data being backed up daily. The team will adhere to established file naming conventions and the project folder structure to ensure consistency and prevent data loss. Data and documents will be archived in their respected format (Excel for raw data and results). Textual data and documents will be saved in PDF/A and plain text (.txt) formats. Datasets will be archived in CSV format for tabular data and JSON or XML for structured data. Codes will be saved on their original document

format. The archived data and documents are going to be stored in the team's Google Drive for remote access. Additionally, copies of the archived data will be stored on encrypted external hard drives kept in secure locations. The team will stick with our file naming conventions and project folder structure.

Professor [Name], the senior member of the research team and a faculty member, will be primarily responsible for the long-term stewardship of the data. Responsibilities will include overseeing the data preservation strategy, ensuring data accessibility, and reviewing data integrity periodically. [Name], as an undergraduate student and coresearcher, will assist with these tasks under Professor [Name]'s guidance.

Finally, we intend to publish our research data to support open scientific inquiry and allow others in the academic community to benefit from our work. The data will be published in the university institutional research repository, adhering to the university's guidelines for data publication. This repository ensures the data is stored securely, made accessible to the academic community, and cited appropriately. We will reach out to our instructors and the repository administrators for guidance on the publication process to ensure compliance with all relevant policies and standards.

5.10. DATA REPOSITORY SELECTION GUIDE

A data repository is a place to archive and make publicly available research datasets. Different teams have different priorities when publishing their data. As such, there are many different data repositories that can be selected to cater to the team, each with its own strengths and capabilities. Data repositories can be divided into two distinct categories:

1. *Generalist data repositories* are data repositories that accept data regardless of the subject, data type, or format.
2. *Domain-specific repositories* are data repositories that limit the data they accept to a specific field or subject.

Your team should analyze your research project to determine the best category for you. The key focus questions are as follows:

TABLE 5.2. *Properties of Common Data Repositories as of Fall 2025*

REPOSITORY NAME	UNLIMITED FILE SIZE?	UNLIMITED DATASET SIZE?	CC0 WAIVER?[1]	DATA AND METADATA SEARCH TOOLS?	DIRECT DOWNLOAD DATA ACCESS?	DATA DOWNLOAD BY API?	TOOLS FOR PROPRIETARY FILE FORMATS?	DATA ANALYSIS TOOLS?	COST?
Dataverse	No	Yes	Advised	Yes	Yes	Yes	No	Yes	None
Dryad	No	Yes	Required	No	Yes	Yes	No	No	Deposition cost
Figshare	No	No	Advised	Yes	Yes	Yes	Yes	No	None
IEEE Dataport	No	No	CC-BY	No	Yes	Yes	Yes	Yes	Deposition + maintenance subscription
Mendeley Data	No	No	Advised	Yes	Yes	Yes	No	No	None
Open Science Framework	No	Yes	Required	Yes	Yes	Yes	No	No	None
Synapse	Yes	Yes	N/A	No	Yes	Yes	No	No	None
Vivli	No	No	Required	Yes	Yes	No	Yes	Yes	Deposition + maintenance subscription
Zenodo	No	No	Required	No	Yes	Yes	No	No	None

1 CC0 (Creative Commons Zero) waives copyright and related rights, placing data in the public domain for unrestricted reuse.

- *Project requirements:* Do your project funders/journals require specific repositories for datasets?
- *Data format limitations:* Does the data repository limit the data and metadata formats you are allowed to use?
- *Costs:* Does the data repository charge a fee for deposition or data storage?
- *Maximum file or database size:* Does the data repository limit the number of bytes each file can have? Does this limit the total size of the overall database?
- *Intellectual property:* Who owns the intellectual property rights? Are there any restrictions or limitations?
- *Sensitive information:* How do you ensure that sensitive data are kept secured, protected, and confidential within the data repository?
- *Integrated tools:* How useful are the tools provided by the data repository for your research? Is there a repository with tools that are better suited for your project?

COMPARING GENERALIST DATA REPOSITORIES

Table 5.2 is a comparison of common generalist repositories (data repositories that accept data regardless of subject, data type, or format). While these are not all of the generalist repositories available, they are some of the most commonly used and may help get a start on finding the repository that is right for your team.

OTHER DATA REPOSITORY SOURCES

If your team needs more information about a data repository or is searching for a repository that focuses on a specific domain or subject, the websites below may be of use.

1. NIH-supported data sharing resources, https://www.nlm.nih.gov/NIHbmic/domain_specific_repositories.html
2. Nature scientific data repository guidance, https://www.nature.com/sdata/policies/repositories

REFERENCES

CITED SOURCES

Beauchamp, T. L., & Childress, J. F. (2009). *Principles of biomedical ethics* (6th ed.). New York, NY: Oxford University Press.

Beever, Jonathan, & Brightman, A. O. (2016). Reflexive principlism as an effective approach for developing ethical reasoning in engineering. *Science and Engineering Ethics* 22 (1): 275–291. https://doi.org/10.1007/s11948-015-9633-5.

Briney, K. (2015). *Data management for researchers: Organize, maintain and share your data for research success.* Pelagic Publishing.

Cornell Data Services. (n.d.). *Writing READMEs for research data.* Accessed January 28, 2026, https://data.research.cornell.edu/data-management/sharing/readme/

Frugoli, J., Etgen, A. M., & Kuhar, M. (2010). Developing and communicating responsible data management policies to trainees and colleagues. *Science and Engineering Ethics, 16*(4), 753–762. https://doi.org/10.1007/s11948-010-9219-1

García-Closas, M., Ahearn, T. U., Gaudet, M. M., et al. (2023). Moving toward findable, accessible, interoperable, reusable practices in epidemiologic research. *American Journal of Epidemiology, 192*(6), 995–1005. https://doi.org/10.1093/aje/kwad040

Gundersen, L. C. (2017). Scientific integrity and ethical considerations for the research data life cycle. In L. C. Gundersen (Ed.), *Scientific integrity and ethics in the geosciences.* American Geophysical Union (AGU). https://doi.org/10.1002/9781119067825.ch9

Jain, S., Pham, T., Morales, J., Rice, A., & Wang, P. (2025). Automated bone fragment classification using computational imaging and machine learning. *The Journal of Purdue Undergraduate Research, 15*, Article 19. https://doi.org/10.7771/2158-4052.1842

Liew, A. (2007). Understanding data, information, knowledge and their inter-relationships. *Journal of Knowledge Management Practice, 8*(2), 1–16.

National Academies of Sciences, Engineering, and Medicine (NASEM). (2018). *Data science for undergraduates: Opportunities and options.* National Academies Press. https://doi.org/10.17226/25104

Obermeyer, Z., Powers, B., Vogeli, C., & Mullainathan, S. (2019). Dissecting racial bias in an algorithm used to manage the health of populations. *Science, 366*(6464), 447–453. https://doi.org/10.1126/science.aax2342

Purzer, Ş., & Arigye, J. (2024). Engineering design reasoning for conscientious decision-making. In L. D. English & T. Lehmann (Eds.), *Ways of thinking in STEM-based problem solving: Teaching and learning in a new era*. Routledge.

Purzer, Ş., Zoltowski, C. B., Zakharov, W., & Arigye, J. (2024). Developing the design reasoning in data life-cycle ethical management framework. *2024 ASEE Annual Conference & Exposition*. https://doi.org/10.18260/1-2--47172

Quintana-Cifuentes, J., & Purzer, Ş. (2022). Semantic fluency in design reasoning. *International Journal of Engineering Education*, *38*(6), 1891–1903.

Rousi, A. M., Boehm, R. I., & Wang, Y. (2024). Data stewardship: case studies from North American, Dutch and Finnish universities. *Journal of Documentation*, 80(7), 306–324. https://doi.org/10.1108/JD-12-2023-0264

Wendelborn, C., Anger, M., & Schickhardt, C. (2023). What is data stewardship? Towards a comprehensive understanding. *Journal of Biomedical Informatics*, *140*(April), Article 104337. https://doi.org/10.1016/j.jbi.2023.104337

Wilkinson, M. D., Dumontier, M., Aalbersberg, Ij. J., Appleton, G., Axton, M., Baak, A., Blomberg, N., Boiten, J.-W., da Silva Santos, L. B., Bourne, P. E., Bouwman, J., Brookes, A. J., Clark, T., Crosas, M., Dillo, I., Dumon, O., Edmunds, S., Evelo, C. T., Finkers, R., … Mons, B. (2016). The FAIR Guiding Principles for scientific data management and stewardship. *Scientific Data*, 3(1), 160018. https://doi.org/10.1038/sdata.2016.18

ADDITIONAL SOURCES

The resources listed in this section complement the material presented throughout the book. They include educational materials, websites, tools, standards, and other references that readers may find useful for extending their knowledge and applying data stewardship principles in practice.

Alipour, J., & Ahmadi, M. (2017). Dimensions and assessment methods of data quality in health information systems. *Acta Medica Mediterranea*, *33*(2), 313–320.

Arigye, J., Zakharov, W., Zoltowski, C., Sewell, S., & Purzer, Ş. (2026). Investigating graduate mentor perspectives on data information literacy and stewardship in undergraduate engineering research projects. *Journal of eScience Librarianship*, *15*(1), e1171. https://doi.org/10.7191/jeslib.1171

Batini, C., Cappiello, C., Francalanci, C., & Maurino, A. (2009). Methodologies for data quality assessment and improvement. *ACM Computing Surveys*, *41*(3), Article 16. https://doi.org/10.1145/1541880.1541883

Bauder, J. (2022). *Teaching research data management*. ALA Editions.

Boehm, R. (n.d.). *Data storage at Purdue: Purdue supported storage*. Purdue Libraries Research Guides. Accessed December 19, 2025, https://guides.lib.purdue.edu/c.php?g=546258&p=5051071

Bradley-Ridout, G. (2018). Preferred but not required: Examining research data management roles in health science librarian positions. *Journal of the Canadian Health Libraries Association / Journal De L Association De Bilbiothèques De La Santé Du Canada, 39*, 138–145. https://doi.org/10.29173/jchla29368

Cai, L., & Zhu, Y. (2015). The challenges of data quality and data quality assessment in the big data era. *Data Science Journal, 14*, Article 2. https://doi.org/10.5334/dsj-2015-002

Cao, L. (2017). Data science: A comprehensive overview. *ACM Computer Survey, 50*(3), Article 43. https://doi.org/10.1145/3076253

Cappiello, C., Francalanci, C., & Pernici, B. (2004). Data quality assessment from the user's perspective. *Proceedings of the ACM SIGMOD International Conference on Management of Data*, pp. 68–73. https://doi.org/10.1145/1012453.1012465

Carlson, J., Fosmire, M., Miller, C.C., & Sapp Nelson, M. (2011). Determining data information literacy needs: A study of students and research faculty. *Portal: Libraries and the Academy, 11*(2), 629–657.

Carlson, J., Sapp Nelson, M., Johnston, L. R., & Koshoffer, A. (2015). Developing data literacy programs: Working with faculty, graduate students and undergraduates. *Bulletin of the Association for Information Science and Technology, 41*(6), 14–17. https://doi.org/10.1002/bult.2015.1720410608

CIO. (2018, February 20). *Data, metadata and the AI horizon*. Accessed December 19, 2025, https://www.cio.com/article/228459/data-metadata-and-the-ai-horizon.html

Coates, H., Carlson, J., Clement, R., Henderson, M., Johnston, L., & Shorish, Y. (2018). How are we measuring up? Evaluating research data services in academic libraries. *Journal of Librarianship and Scholarly Communication, 6*(1), Article eP2226. https://doi.org/10.7710/2162-3309.2226

Conrad, S., Shorish, Y., Whitmire, A., & Hswe, P. (2016). Building professional development opportunities in data services for academic librarians. *IFLA Journal, 43*(1), 65–80. https://doi.org/10.1177/0340035216678237

Cornell Data Services. (n.d.). *Data citation*. Accessed December 19, 2025, https://data.research.cornell.edu/data-management/storing-and-managing/data-citation/

Cornell Data Services. (n.d.). *Writing READMEs for research code & software*. Accessed January 13, 2026, https://data.research.cornell.edu/data-management/sharing/writing-readmes-for-research-code-software/

Cox, A., Kennan, M. A., Lyon, L., & Pinfield, S. (2017). Developments in research data management in academic libraries: Towards an understanding of research data service maturity. *Journal of the Association for Information Science and Technology*, *68*(9), 2182–2200. https://doi.org/10.1002/asi.23781

Cox, A., Kennan, M. A., Lyon, L., Pinfield, S., & Sbaffi, L. (2019). Maturing research data services and the transformation of academic libraries. *Journal of Documentation*, *75*(6), 1432–1462. https://doi.org/10.1108/jd-12-2018-0211

Creative Commons. (n.d.). *About CC licenses*. Accessed December 19, 2025, https://creativecommons.org/share-your-work/cclicenses/

Creative Commons (n.d.). *Licenses list*. Accessed December 19, 2025, https://creativecommons.org/licenses/

DataCite Commons. (n.d.). Accessed December 19, 2025, https://commons.datacite.org/repositories

Data Quality Literacy. (n.d.). *About this project*. Accessed December 19, 2025, https://www.dataqualityliteracy.org/

DDI Alliance. (2025, October 13). *Data Documentation Initiative (DDI)*. https://ddialliance.org.

Dublin Core. (n.d.). *DCMI metadata terms*. Accessed December 19, 2025, https://www.dublincore.org/specifications/dublin-core/dcmi-terms/

Duke University Libraries. (n.d.). *Research data management*. Accessed December 19, 2025. https://library.duke.edu/data/data-management

Evergreen Data. (n.d.). Accessed December 19, 2025, https://stephanieevergreen.com/qualitative-viz/

FAIRsharing. (n.d.). *Databases: A registry of knowledgebases and repositories of data and other digital assets*. Accessed December 19, 2025, https://fairsharing.org/search?fairsharingRegistry=Database

FAIRsharing Team & Pejša, S. (n.d.). *Purdue University Research Repository (PURR)*. https://doi.org/10.25504/FAIRSHARING.785D2F

Fortra. (n.d.). *What is a data repository? (Definition, examples, & tools)*. Accessed December 19, 2025, https://www.fortra.com/blog/what-data-repository

Fox, C., Levitin, A., & Redman, T. (1994). The notion of data and its quality dimensions. *Information Processing & Management*, *30*(1), 9–19. https://doi.org/10.1016/0306-4573(94)90020-5

Getty Research Institute. (n.d.). *Getty thesaurus of geographic names online*. Accessed December 19, 2025, https://www.getty.edu/research/tools/vocabularies/tgn/index.html

Giese, T. G., Wende, M., Bulut, S., & Anderl, R. (2020, April). Introduction of data literacy in the undergraduate engineering curriculum. *2020 IEEE Global*

Engineering Education Conference (EDUCON), pp. 1237–1245. https://doi.org/10.1109/EDUCON45650.2020.9125212

Global Indigenous Data Alliance. (n.d.). *CARE principles for indigenous data governance.* Accessed December 19, 2025, https://www.gida-global.org/care

Go FAIR. (n.d.). *FAIR principles.* Accessed December 19, 2025, https://www.go-fair.org/fair-principles/

Gunsalus, C. K. (2010). Best practices in communicating best practices. *Science and Engineering Ethics, 16*(4), 763–767. https://doi.org/10.1007/s11948-010-9227-1

Harvard Medical School. (n.d.). *Data repositories.* Accessed December 19, 2025, https://datamanagement.hms.harvard.edu/share-publish/data-repositories

Ho, R. C. Y., Wong, S. N., Chia, P., Tang, C., & Ng, M. T. T. (2026). Research data management services in academic libraries to support the research data life cycle: A systematic review. An annual review of information science and technology (ARIST) paper. *Journal of the Association for Information Science and Technology, 77*(1), 272–300. https://doi.org/10.1002/asi.70008

Introduction—Data Management for Undergraduate Researchers—Research Guides at Purdue University. https://guides.lib.purdue.edu/undergraddata

Johnston, L., & Carlson, J. (2015). *Data information literacy librarians: Data and the education of a new generation of researchers.* Purdue University Press.

Joo, S., & Peters, C. (2019). User needs assessment for research data services in a research university. *Journal of Librarianship and Information Science, 52*(3), 633–646. https://doi.org/10.1177/0961000619856073

Joo, S., & Schmidt, G. (2021). Research data services from the perspective of academic librarians. *Digital Library Perspectives, 37*(3), 242–256. https://doi.org/10.1108/dlp 10 2020-0106

Liliendahl, H. G. (2014, May 8). Data quality dimensions and real world alignment. *Liliendahl on Data Quality.* https://liliendahl.com/2014/05/08/data-quality-dimensions-and-real-world-alignment/

Liu, L., & Chi, L. (2002). Evolutional data quality: A theory-specific view. *MIT International Conference on Information Quality*, pp. 292–304.

Linn, M. C., Palmer, E., Baranger, A., et al. (2015). Undergraduate research experiences: Impacts and opportunities. *Science, 347*(6222), 1261757–1261757. https://doi.org/10.1126/science.1261757

Marsh, R. (2005). Drowning in dirty data? It's time to sink or swim: A four-stage methodology for total data quality management. *Journal of Database Marketing & Customer Strategy Management, 12*(2), 105–112. https://doi.org/10.1057/palgrave.dbm.3240247

Maybee, C., & Zilinski, L. (2015). Data informed learning: A next phase data literacy framework for higher education. *Proceedings of the Association for Information Science and Technology*, 52(1), 1–4. https://doi.org/10.1002/pra2.2015.1450520100108

Medical Library Association. (2024, December 12). *Good data supports good decisions: A practical guide to data quality checking*. https://www.mlanet.org/courses/good-data-supports-good-decisions-a-practical-guide-to-data-quality-checking/

Michigan State University. *How to cite data: General info*. MSU Libraries Research Guides. Accessed December 19, 2025, https://libguides.lib.msu.edu/c.php?g=96245&p=626236

Miliate, B. (n.d.). *Research data management: Overview*. Yale University Library Research Guides. Accessed December 19, 2025, https://guides.library.yale.edu/datamanagement/home

National Information Standards Organization (NISO). (n.d.). *The importance of metadata in AI* [Webinar]. Accessed December 19, 2025, https://www.niso.org/events/importance-metadata-ai

Nature. (n.d.). *Scientific data: Data repository guidance*. Accessed January 13, 2026, https://www.nature.com/sdata/policies/repositories

Nicholson, R. (2025, October 15). 18 best types of charts and graphs for data visualization [+ how to choose]. *HubSpot*. https://blog.hubspot.com/marketing/types-of-graphs-for-data-visualization.

Nitecki, D. A., & Davis, M. E. K. (2019). Expanding academic librarians' roles in the research life cycle. *Libri*, *69*(2). https://doi.org/10.1515/libri-2018-0066

Open Data Handbook. (n.d.). *What is open data?* Accessed December 19, 2025, https://opendatahandbook.org/guide/en/what-is-open-data/

Open Knowledge. (n.d.). *Open data commons: Legal tools for open data*. Accessed December 19, 2025, https://opendatacommons.org/licenses/

ORCID. Accessed December 19, 2025, https://orcid.org/

Panahy, P. H. S., Sidi, F., Affendey, L. S., & Jabar, M. A. (2014). The impact of data quality dimensions on business process improvement. *4th World Congress on Information and Communication Technologies, WICT 2014*, pp. 70–73. https://doi.org/10.1109/WICT.2014.7077304

Pipino, L. L., Lee, Y. W., & Wang, R. Y. (2002). Data quality assessment. *Communications of the ACM*, *45*(4), 211–218. https://doi.org/10.1145/505248.506010

Purdue OWL. (n.d.). *General format—APA citations*. Accessed December 19, 2025, https://owl.purdue.edu/owl/research_and_citation/apa_style/apa_formatting_and_style_guide/general_format.html

Ray, J. M. (Ed.). (2014). *Research data management: Practical strategies for information professionals*. DGO—Digital original. Purdue University Press. https://doi.org/10.2307/j.ctt6wq34t

Registry of Research Data Repositories. (n.d.) https://www.re3data.org/

Riley, J. (n.d.). *Understanding metadata: What is metadata, and what is it for?* National Information Standards Organization. Accessed December 19, 2025, https://www.niso.org/publications/understanding-metadata-2017

Rod, A., Hervieux, S., & Lee, N. (2024). Evaluating an instructional intervention for research data management training. *Evidence Based Library and Information Practice, 19*(1), 114–131. https://doi.org/10.18438/eblip30439

Rousi, A. M., Boehm, R. I., & Wang, Y. (2024). Data stewardship: case studies from North American, Dutch and Finnish universities. *Journal of Documentation, 80*(7), 306–324. https://doi.org/10.1108/JD-12-2023-0264

Sare, L. (n.d.). *Research data management: Overview.* Texas A&M University Libraries Research Guides. Accessed December 19, 2025, https://tamu.libguides.com/research-data-management/overview

ScienceDirect. (n.d.). *Descriptive metadata.* Accessed December 19, 2025, https://www.sciencedirect.com/topics/computer-science/descriptive-metadata

Scientific Data. (n.d.). *Data repository guidance.* Accessed December 19, 2025, https://www.nature.com/sdata/policies/repositories

Seagate. (n.d.). What is a 3-2-1 backup strategy? *Seagate Blog.* Accessed December 19, 2025, https://www.seagate.com/blog/what-is-a-3-2-1-backup-strategy/

Sidi, F., Shariat Panahy, P. H., Affendey, L. S., Jabar, M. A., Ibrahim, H., & Mustapha, A. (2012). Data quality: A survey of data quality dimensions. *2012 International Conference on Information Retrieval and Knowledge Management, CAMP'12*, pp. 300–304. https://doi.org/10.1109/InfRKM.2012.6204995

Singers, M. (n.d.). *Awesome README.* https://github.com/matiassingers/awesome-readme

SPARC. (n.d.). *Open data.* Accessed December 19, 2025, https://sparcopen.org/open-data/

Strasser, C. A., & Hampton, S. E. (2012). The fractured lab notebook: Undergraduates and ecological data management training in the United States. *Ecosphere, 3*(12), Article 116. https://doi.org/10.1890/ES12-00139.1

Tang, R., & Hu, Z. (2019). Providing research data management (RDM) services in libraries: Preparedness, roles, challenges, and training for RDM practice. *Data and Information Management, 3*(2), 84–101. https://doi.org/10.2478/dim-2019-0009

TDWG. (n.d.). Darwin Core main vocabulary. Accessed December 19, 2025, https://rs.tdwg.org/dwc.htm

Timms, G. (2024). Introducing undergraduates to research data management: A marine science case study-based asynchronous express course. *Science & Technology Libraries, 43*(1), 78–97. https://doi.org/10.1080/0194262X.2023.2237077

University of Nebraska–Lincoln Libraries. (n.d.). *Research data management & sharing.* Accessed December 19, 2025, https://libraries.unl.edu/research-data-management/

University of Pittsburgh Library System. (n.d.). *Data management and digital project management support.* Accessed December 19, 2025, https://library.pitt.edu/data-management

University of Texas Libraries. (n.d.). *Metadata basics: Standards.* Accessed December 19, 2025, https://guides.lib.utexas.edu/metadata-basics/standards

U.S. Department of Health and Human Services. (n.d.). *Finding datasets, data repositories, and data standards.* https://www.nihlibrary.nih.gov/finding-datasets-data-repositories-and-data-standards

U.S. National Library of Medicine. (n.d.). *Medical subject headings (MeSH).* Accessed December 19, 2025, https://meshb.nlm.nih.gov/#/fieldSearch

U.S. National Library of Medicine. (n.d.). *NIH-supported data sharing resources.* Accessed December 19, 2025, https://www.nlm.nih.gov/NIHbmic/domain_specific_repositories.html

U.S. National Science Foundation. (n.d.). *Open data at NSF.* Accessed December 19, 2025, https://www.nsf.gov/digital/data

U.S. National Science Foundation. (n.d.). *Preparing your data management and sharing plan—funding at NSF.* Accessed December 19, 2025, https://www.nsf.gov/funding/data-management-plan

U.S. National Science Foundation. (2018, November). *ENG data management and sharing plan guidance.* Accessed December 19, 2025, https://www.nsf.gov/eng/data-management-plans

U.S. National Science Foundation. (2023, July). *EAR data management and sharing plan guidance.* Accessed December 19, 2025, https://www.nsf.gov/geo/ear/data-management-sharing-plans

The White House. (2022, August 25). *OSTP issues guidance to make federally funded research freely available without delay.* https://bidenwhitehouse.archives.gov/ostp/news-updates/2022/08/25/ostp-issues-guidance-to-make-federally-funded-research-freely-available-without-delay/

Wiljes, C., & Cimiano, P. (2019). Teaching research data management for students. *Data Science Journal, 18,* Article 38. https://doi.org/10.5334/dsj-2019-038

Xu, Z. (2022). Research data management practice in academic libraries. *Journal of Librarianship and Scholarly Communication, 10*(1). https://doi.org/10.31274/jlsc.13700

Zakharov, W., Purzer, Ş., & Zoltowski, C. B. (2025). Cultivating ethical data stewardship for undergraduate research: A framework and toolkit duo. *Information Outlook,* (Fall), 17–20. https://docs.lib.purdue.edu/lib_fsdocs/329/

Zakharov, W. (n.d.). *Data quality checking guide.* Accessed December 19, 2025, https://docs.lib.purdue.edu/lib_fsdocs/308/

Zakharov, W. (n.d.). *Data management for undergraduate researchers: Data management plans.* Purdue Libraries Research Guides. Accessed December 19, 2025, https://guides.lib.purdue.edu/undergraddata/dmp

Zakharov, W. (n.d.). *Research Guides: Data Stewardship Education Workshop: Home.* Retrieved July 17, 2026, from https://guides.lib.purdue.edu/c.php?g=1536366&p=11516524

ABOUT THE AUTHORS

GRANT PROJECT DESCRIPTION

This book is one of the deliverables of the National Science Foundation Improving Undergraduate STEM Education (IUSE)–funded project, Prioritizing Data Life Cycle Management for Shaping Next Generation Researchers (#2236241).

PROJECT TEAM MEMBERS

FACULTY MEMBERS

ŞENAY PURZER is a professor of engineering education at Purdue University. A Fulbright Specialist, she is recognized for her work in undergraduate engineering education. Purzer is renowned for her expertise in conscientious design reasoning. She has co-led numerous research projects over the past 17 years at Purdue University to improve undergraduate education, including the NSF-funded I project, Prioritizing Data Life Cycle Management for Shaping Next Generation Researchers, that informed the publication of this book.

WEI ZAKHAROV is an associate professor and engineering information specialist in the Libraries and School of Information Studies and (by courtesy) the School of Engineering Education at Purdue University. Her work focuses on data and information literacy education and online learning. She has developed and taught several undergraduate courses, including Foundations of Data Management and Understanding Your Research Data, which supports Purdue's data science and undergraduate research certificates. Zakharov mentors undergraduate research courses and leads National Science Foundation Improving Undergraduate STEM Education (IUSE) and Institute of Museum and Library Services (IMLS) National Leadership Grant projects in an effort to advance undergraduate data stewardship education.

CARLA B. ZOLTOWSKI is an associate professor of engineering practice in the Elmore Family School of Electrical and Computer Engineering (ECE) and (by courtesy) the School of Engineering Education at Purdue University. She is also the director of Purdue's Vertically Integrated Projects (VIP) program and interim director of EPICS. She has mentored hundreds of student design and research teams in EPICS and VIP. Her research includes the professional formation of engineers, human-centered design, and engineering ethics. She is coprincipal investigator on the NSF-funded Improving Undergraduate STEM Education (IUSE) project Prioritizing Data Life Cycle Management for Shaping Next Generation Researchers.

GRADUATE RESEARCH ASSISTANTS

JOREEN ARIGYE, School of Engineering Education, Purdue University (graduated August 2025); assistant professor of practice, Center for Computing Education, Department of Engineering Education, Ohio State University (current)

NIKI FOSTER, School of Engineering Education, Purdue University

UNDERGRADUATE RESEARCHERS

JACOB MORALES, undergraduate researcher, Elmore Family School of Electrical and Computer Engineering, Purdue University

ALLIE RICE, undergraduate researcher, School of Mechanical Engineering, Purdue University

JUAN JOSE RUIZ MAZA NAVA, undergraduate researcher, Edwardson School of Industrial Engineering, Purdue University

SARAH SEWELL, undergraduate researcher and graphic designer, Elmore Family School of Electrical and Computer Engineering, Purdue University

BOOK CASE CONTRIBUTORS

NICOLE ADAMS, associate research professor, Regenstrief Center for Healthcare Engineering, Purdue University

NATHAN JAMES ARNOLD, undergraduate researcher, School of Aeronautics and Astronautics, Purdue University

PARTH KAILASH DUBAL, undergraduate researcher, School of Aeronautics and Astronautics, Purdue University

JAMES MICHAEL GOPPERT, research assistant professor, School of Aeronautics and Astronautics, Purdue University

NAN KONG, professor, School of Biomedical Engineering, Purdue University

VINEETH NARRA, undergraduate researcher, Department of Psychological Sciences, Purdue University

PETE E. PASCUZZI, associate professor, Libraries and School of Information Studies, Purdue University

COLIN ROBERSON, undergraduate researcher, Department of Computer Sciences, Purdue University

LONNIE W. SCHWARTZ, undergraduate researcher, Department of Biochemistry, Purdue University

KARTHIK SELVARAJ, undergraduate researcher, Elmore Family School of Electrical and Computer Engineering, Purdue University

BOOK TOOLKIT REVIEWERS

KRISTIN BRINEY, biology & biological engineering librarian, California Institute of Technology

MICHAEL FOSMIRE, professor, Libraries and School of Information Studies, Purdue University

HAOYONG LAN, engineering librarian, Carnegie Mellon University

MEGAN SAPP NELSON, professor and head, Grainger Engineering Library Information Center, University of Illinois Urbana-Champaign

ZHIHONG XU, associate professor, Department of Agricultural Leadership, Education, and Communications, Texas A&M University

www.ingramcontent.com/pod-product-compliance
Lightning Source LLC
LaVergne TN
LVHW010951010926
846999LV00031B/364
* 9 7 8 1 6 2 6 7 1 3 9 0 1 *